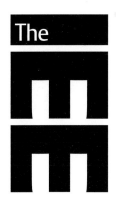

The IEE

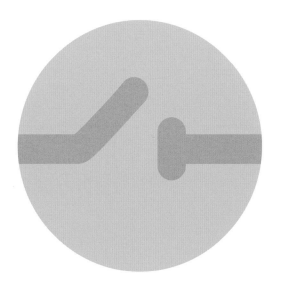

Guidance Note 2
Isolation
& Switching

17th IEE Wiring Regulations Seventeenth Edition
BS 7671:2008 Requirements for Electrical Installations

Published by The Institution of Engineering and Technology, London, United Kingdom

The Institution of Engineering and Technology is registered as a Charity in England & Wales (no. 211014) and Scotland (no. SC038698).

The Institution of Engineering and Technology is the new institution formed by the joining together of the IEE (The Institution of Electrical Engineers) and the IIE (The Institution of Incorporated Engineers). The new Institution is the inheritor of the IEE brand and all its products and services, such as this one, which we hope you will find useful. The IEE is a registered trademark of the Institution of Engineering and Technology.

First published 1992 (0 85296 536 2)
Reprinted (with minor amendments) 1993
Second edition (incorporating Amendment No. 1 to BS 7671:1992) 1995 (0 85296 866 3)
Third edition (incorporating Amendment No. 2 to BS 7671:1992) 1999 (0 85296 955 4)
Fourth edition (incorporating Amendment No. 1 to BS 7671:2001) 2003 (0 85296 990 2)
Reprinted (incorporating Amendment No. 2 to BS 7671:2001) 2004
Fifth edition (incorporating BS 7671:2008) 2009 (978-0-86341-856-3)

Copies of this publication may be obtained from:
The Institution of Engineering and Technology
PO Box 96
Stevenage
SG1 2SD, UK
Tel: +44 (0)1438 767328
Email: sales@theiet.org
www.theiet.org/publishing/books/wir-reg/

ISBN 978-0-86341-856-3

Typeset in the UK by Carnegie Book Production, Lancaster
Printed in the UK by Printwright Ltd, Ipswich

Contents

Cooperating organisations

The Institution of Engineering and Technology acknowledges the contribution made by the following organisations in the preparation of this Guidance Note.

British Electrotechnical & Allied Manufacturers Association Ltd
P. Galbraith IEng MIET MCMI

BEAMA Installation
Eur Ing M.H. Mullins BA CEng FIEE FIIE
P. Sayer IEng MIET GCGI

Electrical Contractors' Association (ECA)
Eur Ing L. Markwell MSc BSc CEng MIET MCIBSE LCGI

GAMBICA Association Ltd
P. Still MIEE

Health and Safety Executive (HSE)
N. Gove MEng CEng MIEE
K. Morton BSc CEng MIEE

Institution of Engineering and Technology
P. Bicheno BSc(Hons) MIET
M. Coles BEng(Hons) MIET
G. Cronshaw IEng FIET
P.E. Donnachie BSc CEng FIET
J. Elliott BSc(Hons) PG Cert IEng MIET (Editor)

PLASA
J. Eade

Acknowledgements

References to British Standards, CENELEC Harmonization Documents and International Electrotechnical Commission standards are made with the kind permission of BSI. Complete copies can be obtained by post from:

BSI Customer Services
389 Chiswick High Road
London W4 4AL
Tel: +44 (0)20 8996 9001
Email: orders@bsi-global.com

BSI also maintains stocks of international and foreign standards, with many English translations. Up-to-date information on BSI standards can be obtained from the BSI website: www.bsi-global.com

Advice is included with the kind permission of the Energy Networks Association Limited. Complete copies of their publications can be obtained by post from:

Energy Networks Association
6th Floor, Dean Bradley House
52 Horseferry Road
London SW1P 2AF
Tel: +44 (0)20 7706 5100
Email: info@energynetworks.org

Documents available from their website www.energynetworks.org include Technical Specifications, BEBS Specifications, Engineering Recommendations and a variety of reports.

Copies of Health and Safety Executive documents and approved codes of practice (ACOP) can be obtained from:

HSE Books
PO Box 1999
Sudbury, Suffolk CO10 2WA
Tel: +44 (0)1787 881165
Email: hsebooks@prolog.uk.com

The HSE website is www.hse.gov.uk

The illustrations within this publication were provided by Rod Farquhar Design: www.rodfarquhar.co.uk

Cover design and illustration were created by The Page Design: www.thepagedesign.co.uk

Preface

This Guidance Note is part of a series issued by the Institution of Engineering and Technology to explain and enlarge upon the requirements in BS 7671:2008, the 17th Edition of the IEE Wiring Regulations.

Note that this Guidance Note does not ensure compliance with BS 7671. It is intended to explain some of the requirements of BS 7671, but readers should always consult BS 7671 to satisfy themselves of compliance.

The scope generally follows that of BS 7671; the relevant Regulations and Appendices are noted in the margin. Some Guidance Notes also contain material not included in BS 7671:2008 but which was included in earlier editions of the Wiring Regulations. All of the Guidance Notes contain references to other relevant sources of information.

Electrical installations in the United Kingdom that comply with BS 7671 are likely to satisfy Statutory Regulations such as the Electricity at Work Regulations 1989; however, this cannot be guaranteed. It is stressed that it is essential to establish which Statutory and other Regulations apply and to install accordingly. For example, an installation in premises subject to licensing may have requirements different from, or additional to, BS 7671 and these will take precedence.

Introduction

This Guidance Note is concerned primarily with Section 537 and those other parts of BS 7671 which relate to isolation, switching and control including provision of information and labelling. Requirements relating to isolation and switching in particular types of installation and for specific items of installed equipment are also covered.

Neither BS 7671 nor the Guidance Notes are design guides. It is essential to prepare a full design and specification prior to commencement or alteration of an electrical installation. Compliance with the relevant standards should be required.

The design and specification should set out the requirements and provide sufficient information to enable competent persons to carry out the installation and to commission it. The specification must include a description of how the system is to operate and all the design and operational parameters. It must provide for all the commissioning procedures that will be required and for the provision of adequate information to the user. This will be by means of an operational and maintenance manual or schedule, and 'as fitted' drawings if necessary. **514.9**

It must be noted that it is a matter of contract as to which person or organisation is responsible for the production of the parts of the design, specification, construction and verification of the installation and any operational information.

The persons or organisations who may be concerned in the preparation of the works include:

> The Designer
> The CDM Coordinator
> The Installer
> The Supplier of Electricity (Distributor)
> The Installation Owner (Client) and/or User
> The Architect
> The Fire Prevention Officer
> All Regulatory Authorities
> Any Licensing Authority
> The Health and Safety Executive

In producing the design, advice should be sought from the installation owner and/or user as to the intended use. Often, as in a speculative building, the intended use is unknown. The specification and/or the operational manual must set out the basis of use for which the installation is suitable. **132.1**

Precise details of each item of equipment should be obtained from the manufacturer and/or supplier and compliance with appropriate standards confirmed. **133.1** **Section 511**

The operational manual must include a description of how the system as installed is to operate and all commissioning records. The manual should also include manufacturers' technical data for all items of switchgear, luminaires, accessories, etc. and any special instructions that may be needed.

The Health and Safety at Work etc. Act 1974 Section 6 and the Construction (Design and Management) Regulations 2007 are concerned with the provision of information, and guidance on the preparation of technical manuals is given in the BS 4884 series *Technical manuals* and the BS 4940 series *Technical information on constructional products and services*. The size and complexity of the installation will dictate the nature and extent of the manual.

Statutory requirements

1

1.1 General

A number of enactments and statutory instruments including the Electricity at Work Regulations 1989 (made under the Health and Safety at Work etc. Act 1974) are of relevance to isolation and switching. These include machinery safety requirements which also come under the Supply of Machinery (Safety) Regulations 1992 as amended.

This Guidance Note is not intended to provide an exhaustive treatment of the legislation concerned with isolation and switching in low voltage installations, but deals only with those situations referred to in BS 7671. Thus, certain specialised installations listed in Regulation 110.2 are excluded.

110.2

1.2 Statutory regulations

Appx 2

1.2.1 The Health and Safety at Work etc. Act 1974

The Health and Safety at Work etc. Act 1974, Part I, Section 6, places a duty on any person who designs, manufactures, imports or supplies any article for use at work to ensure that adequate information is provided so that when put to that use, it will be safe and without foreseeable risks to health. Schedule 3 of the Consumer Protection Act 1987 extended the use to when the equipment is being set, used, cleaned or maintained.

Those with duties under the Act should include in their written instructions, manuals, etc. details of the means of isolation and other health and safety provisions provided and the need to use these when and where required.

An electrical installation may well, under some circumstances, be considered to be 'an article for use at work' and there is a duty on the designer and installer to provide adequate instruction and/or maintenance manuals irrespective of any contract provisions.

1.2.2 The Electricity at Work Regulations 1989 as amended

The Electricity at Work Regulations (EWR) 1989 are general in their application and refer throughout to 'danger' and 'injury'. Danger is defined as risk of 'injury' and injury is defined in terms of certain classes of potential harm to persons. Injury is stated to mean death or injury to persons from:

▶ electric shock
▶ electric burn
▶ electrical explosion or arcing
▶ fire or explosion initiated by electrical energy.

Of particular interest to the subject matter of this publication are Regulation 12 'Means for cutting off the supply and for isolation' and Regulation 13 'Precautions for work on equipment made dead'.

Regulation 12(1)(b) states that where necessary to prevent danger, suitable means shall be available for the isolation of any electrical equipment, where isolation means the disconnection and separation of the electrical equipment from every source in such a way that the disconnection and separation is secure (para (2) refers).

The *Memorandum of guidance on the Electricity at Work Regulations 1989* (HSR25) published by the Health and Safety Executive advises with reference to Regulation 12(1)(b) above that isolation is the process of ensuring that the supply to all or a particular part of an installation remains switched off and that inadvertent reconnection is prevented.

The issue of preventing inadvertent reconnection is covered in Regulation 13 of the EWR 1989:

> ### 13. Precautions for work on equipment made dead
>
> Adequate precautions shall be taken to prevent electrical equipment , which has been made dead in order to prevent danger while work is being carried out on or near that equipment, from becoming electrically charged during that work if danger may thereby arise.

The coverage of Regulation 13 in HSR25 highlights the need to lock off or otherwise secure any switching device being used to provide isolation, or where isolation has been achieved by the removal of fuses, by their retention in a safe place during the isolation period.

The *Memorandum of guidance on the Electricity at Work Regulations 1989* (HSR25) is essential reading for all concerned with electrical installations.

BS 7671 is intended primarily for designers and installers and thus covers only the provision of isolators and the means of securing them. The responsibility for ensuring that equipment is properly isolated when necessary remains with the user.

HSR25 In the above-mentioned Memorandum (see paragraph 182 relating to Regulation 12(1)(a)) it is recognised that there may be a need to switch off electrical equipment for reasons other than preventing electrical danger, but these considerations are outside the scope of the Electricity at Work Regulations 1989.

131.1(v) BS 7671, however, takes into account injury from mechanical movement of electrically actuated equipment. Emergency switching and switching off for mechanical maintenance are therefore included in BS 7671.

The Electricity at Work Regulations 1989 specifically require adequate maintenance, and this implies inspection of electrical systems, supplemented by testing as necessary. Regular functional testing of safety circuits such as emergency switching/stopping, etc. may be required — especially if they are very rarely required to be used. Comprehensive records of all inspections and tests should be made and reviewed for any trends that GN3 may arise. Guidance Note 3: *Inspection & Testing* gives more detailed guidance on initial and periodic inspection and testing of installations.

1.2.3 The Electricity Safety, Quality and Continuity Regulations 2002 as amended

The prime purpose of the Electricity Safety, Quality and Continuity Regulations (ESQCR) 2002 is to provide for the safety of the public and to ensure an electricity supply of adequate quality and reliability. These Regulations make reference to BS 7671. For example, under Regulation 25(2) the consumer may have to satisfy the local distributor that the electrical installation is safe and technically sound by providing evidence that it complies with BS 7671. An Electrical Installation Certificate would normally be acceptable.

1.2.4 The Machinery Directive

The EU Machinery Directive was implemented into UK law by the Supply of Machinery (Safety) Regulations 1992 (SI 1992 No. 3073 made under the European Communities Act 1972). This Directive will be superseded on 29 December 2009 by the new Machinery Directive, 2006/42/EC. The new Directive aims to provide greater clarity than the old directive, e.g. with a modified definition of the core concept of 'machinery' and in the dividing lines between itself and the Lifts and Low Voltage Directives. Another significant change is the introduction of a quality assurance module as a conformity assessment option for relevant manufacturers.

At the time of writing, the changes to the UK Regulations resulting from the new Directive have not been determined.

These Regulations, amended in 1994, have specific requirements for the essential health and safety aspects of machines. The requirements are wide ranging and take into account potential dangers to operators of machinery and to other persons within a 'danger zone'. Only the requirements for isolation, switching and control are relevant to this Guidance Note.

In the Supply of Machinery (Safety) Regulations, machinery is defined as:

(a) an assembly of linked parts or components, at least one of which moves with the appropriate actuators, control and power circuits, joined together for a specific application, in particular for the processing, treatment, moving or packaging of a material;

(b) an assembly of machines which, in order to achieve the same end, are arranged and controlled so that they function as an integral whole; or

(c) interchangeable equipment modifying the function of a machine which is supplied for the purpose of being assembled with an item of machinery or with a series of different items of machinery or with a tractor by the operator himself save for any such equipment which is a spare part or tool.

The Regulations do not apply:

i   to machinery specific to specialist equipment listed in Schedule 5 of the Regulations, including ships and offshore platforms for which reference should be made to the relevant statutory instrument

ii to machinery previously used in the European Community

iii to machinery for use outside the European Economic Area

iv where the risks are mainly of electrical origin (such machinery is covered by the Electrical Equipment (Safety) Regulations 1994, SI 1994 No. 3260) – see section 1.2.5. Note: These Statutory Regulations apply only to low voltage equipment up to 1 kV

v where the risks are wholly or partly covered by other Directives, from the date those other Directives are implemented into United Kingdom law

vi to machinery first supplied in the European Community before 1 January 1993.

Machinery manufactured in conformity with specified published European standards that have also been listed in the *Official Journal of the European Communities* will be presumed to comply with the essential health and safety requirements of those standards and hence the Regulations.

BS EN 60204
BS EN ISO 12100
BS EN ISO 13850

BS EN 60204 *Safety of Machinery: Electrical equipment of machines* is the major standard on machine electrical equipment. BS EN ISO 12100 *Safety of machinery. Basic concepts, general principles for design* and BS EN ISO 13850 *Safety of machinery. Emergency stop equipment, functional aspects. Principles for design* also give advice on emergency switching.

The principles of safety stated in Schedule 3 of the Supply of Machinery (Safety) Regulations 1992 as amended are:

1.1.2 Principles of safety integration

(a) Machinery must be so constructed that it is fit for its function and can be adjusted and maintained without putting persons at risk when these operations are carried out under the conditions foreseen by the manufacturer.
The aim of the measures taken must be to minimise foreseeable risks of accident throughout the intended lifetime of the machinery, including the phases of assembly and dismantling, even where risks of accident arise from foreseeable abnormal situations.

(b) In selecting the most appropriate methods, the manufacturer must apply the following principles, in the order given:
– eliminate or reduce risks as far as practicable by inherently safe machinery design and construction;
– take the necessary protection measures in relation to risks that cannot be eliminated;
– inform users of the residual risks due to any shortcomings of the protection measures adopted, indicate whether any particular training is required and specify any need to provide personal protection equipment.

(c) When designing and constructing machinery and when drafting the instructions, the manufacturer must envisage not only the normal use of the machinery but also other uses which could reasonably be foreseen.
The machinery must be designed to prevent abnormal use if such use would engender a risk.
The instructions must also draw the user's attention to ways — that experience has shown might occur — in which the machinery should not be used.

(d) Under the intended conditions of use, the discomfort, fatigue and psychological stress faced by the operator must be reduced to the minimum possible taking ergonomic principles into account.

(e) When designing and constructing machinery, the manufacturer must take account of the constraints to which the operator is subject as a result of the necessary or foreseeable use of personal protection equipment such as footwear, gloves, etc.

(f) Machinery must be supplied with all the essential special equipment and accessories to enable it to be adjusted, maintained and used without risk.

1.2.5 The Electrical Equipment (Safety) Regulations 1994

As noted in section 1.2.4, certain machines are excluded from the Supply of Machinery (Safety) Regulations, where these have risks that are mainly of electrical origin. Such equipment is covered by the Electrical Equipment (Safety) Regulations 1994. These apply to equipment covered by the EU area Low Voltage Directive (LVD). Low voltage is defined in Part 2 of BS 7671 as not exceeding 1000 V a.c. or 1500 V d.c.

1.2.6 The Construction (Design and Management) Regulations 2007

The Construction (Design and Management) Regulations (CDM Regulations) require active planning, coordination and management of the building works, including the electrical installation, to ensure that hazards associated with the construction, maintenance and perhaps demolition of the installation are given due consideration, as well as provision for safety in normal use.

Regulation 20(2) requires a record known as 'the health and safety file' to be prepared, reviewed and updated during the construction process. This file should be passed to the client on completion of the construction work.

Regulation 17(3) requires that reasonable steps are taken to ensure that once the construction phase has been completed the information in the health and safety file remains available for inspection by any person who might need it to comply with any relevant statutory provisions. It also requires that the file is revised and updated as often as may be appropriate to incorporate any relevant new information.

1.2.7 The Management of Health and Safety at Work Regulations 1999

These Regulations place general duties on employers to assess risks to the health and safety of employees and others and take managerial action to minimise these risks, including:

▶ implementing preventive measures
▶ providing health surveillance
▶ appointing competent people
▶ setting up procedures
▶ providing information
▶ training, etc.

By its very nature the risk assessment must include operation of equipment and machines, and the safety of the fixed installation, and must cover the adequate provision of isolation and emergency switching and stopping devices suitable for the considered risk. It should also cover access interlocks and any other equipment safety operating provisions such as guards, etc.

The general requirement is to identify and design to reduce risks to protect against hazards and dangers.

Note: These Regulations have been subject to some revocations under the Regulatory Reform (Fire Safety) Order 2005, which is described in section 1.2.9. See Schedule 5 of the Order for full details.

1.2.8 The Provision and Use of Work Equipment Regulations 1998

These Regulations require employers to ensure that work equipment is suitable for the purpose.

The Regulations also require that work equipment is maintained in an efficient state, in efficient working order and in good repair.

Work equipment is defined in the Regulations as being any machine, equipment or tool or installation used at work, whether exclusively or not.

Regulations 14 to 18 cover controls and control systems, Regulation 19 details isolation requirements and Regulation 21 requires the provision of suitable and sufficient lighting for safe operation and working. Regulations 5, 6 and 22 are concerned with equipment maintenance.

1.2.9 The Regulatory Reform (Fire Safety) Order 2005

This is probably the single most influential statutory instrument in the field of protection against fire. It has a direct influence on other pieces of primary and secondary statutory legislation, requiring modifications to and in some cases partial or full revocation of the requirements therein. It replaces fire certification under the Fire Precautions Act 1971 with a general duty to ensure, so far as is reasonably practicable, the safety of employees and a general duty, in relation to non-employees, to take such fire precautions as may reasonably be required to ensure that premises are safe. It also requires a risk assessment to be carried out and regularly updated.

The Regulatory Reform (Fire Safety) Order 2005 (henceforth referred to as the Order) came into effect fully in October 2006 and affects over 70 pieces of fire safety law, not all of which fall within the scope of this publication.

The Order applies to non-domestic premises in England and Wales, including any communal areas of blocks of flats or houses in multiple occupation, and places legal obligations and responsibilities upon the *responsible person* defined in article 3 of the Order as follows:

> **3. Meaning of 'responsible person'**
>
> In this Order 'responsible person' means –
>
> (a) in relation to a workplace, the employer, if the workplace is to any extent under his control;
> (b) in relation to any premises not falling within paragraph (a) –
>   (i) the person who has control of the premises (as occupier or otherwise) in connection with the carrying on by him of a trade, business or other undertaking (for profit or not); or
>   (ii) the owner, where the person in control of the premises does not have control in connection with the carrying on by that person of a trade, business or other undertaking.

Part 2 of the Order 'Fire Safety Duties' requires the responsible person to take the necessary general fire precautions to ensure, so far as is reasonably practicable, the safety of employees and other persons within the premises for which they are responsible. The responsible person must carry out a fire safety risk assessment on the premises to identify what general fire precautions are required for the above to be achieved. The risk assessment must be regularly reviewed by the responsible person to keep it up to date.

The significant findings of the risk assessment including any actions that have been taken or which will be taken by the responsible person and details of any persons identified as being especially at risk must be recorded as soon as practicable afterwards where:

▶ the responsible person employs five or more persons, or
▶ the premises to which the assessment relates are subject to a licensing arrangement, or
▶ the premises are subject to an alterations notice.

Article 23 requires that every employee must, while at work:

(a) take reasonable care for the safety of themselves and of other persons who may be affected by their acts or omissions at work;
(b) as regards any duty or requirement imposed on their employer by or under any provision of the Order, cooperate with them so far as is necessary to enable that duty or requirement to be performed or complied with; and
(c) inform their employer or any other employee with specific responsibility for the safety of fellow employees –
   (i) of any work situation which they would reasonably consider represented a serious and immediate danger to safety; and
   (ii) of any matter which they would reasonably consider represented a shortcoming in the employer's protection arrangements for safety,
   in so far as that situation or matter either affects the safety of the employee or arises out of or in connection with their activities at work, and has not previously been reported to the employer.

Article 37 relates to firefighters' switches for luminous tube signs, etc. This is a topic dealt with in greater depth in Chapter 8 of this publication.

Reference should be made to the various schedules to the Order, which are as follows:

Schedule 1   This consists of four parts. Parts 1 and 2 relate to risk assessment, Part 3 relates to principles of prevention and Part 4 relates to measures to be taken in respect of dangerous substances.
Schedule 2   Amendments of primary legislation
Schedule 3   Amendments of subordinate legislation
Schedule 4   Repeals
Schedule 5   Revocations

It should be noted that in Scotland the Fire (Scotland) Act 2005 provides similar requirements for fire safety duties to those of the Regulatory Reform (Fire Safety) Order 2005 applicable to England and Wales.

Guidance Note 2: Isolation & Switching
© The Institution of Engineering and Technology

Overview of the Wiring Regulations

2

Chapter 13 of BS 7671 prescribes the fundamental principles for safety and includes a number of regulations which are directly concerned with isolation and switching.

Chap 13

In particular, Regulation 131.7 requires consideration to be given to the effects on the installation and equipment installed therein caused by an interruption of the supply. Regulation 132.9 states the fundamental requirement that any emergency control devices be so installed as to be clearly recognisable and effectively and rapidly operated in the case of danger arising. Regulation 132.10 requires the suitable provision of disconnecting devices to permit the safe use of the installation in terms of operation; inspection; fault detection, diagnosis and repair; routine testing; and maintenance. Although not specific to isolation and switching alone, the accessibility requirements of Regulation 132.12 should be borne in mind when considering suitable location for items of switchgear within an installation. This is also true of the requirements of Regulation 132.13 regarding the provision of suitable documentation for the installation.

131.7

132.9

132.10

132.12

132.13

Regulation 132.14.1 makes it clear that single-pole devices including fuses, switches and circuit-breakers may only be installed in the line conductor of a circuit, while Regulation 132.14.2 prohibits the installation of such single-pole devices in an earthed neutral conductor.

132.14.1

132.14.2

Regulation 132.15.1 requires the provision of a readily accessible means of isolation where necessary to prevent or remove danger, at the origin of an installation, at the beginning of each circuit and where necessary to individual items of equipment. Regulation 132.15.2 requires the provision of a readily accessible means of switching off for every motor installed as fixed equipment. This requirement is frequently misunderstood to mean that a means of local isolation must be provided for every motor in an installation, which is not the case.

132.15.1

132.15.2

Regulation group 133.1 requires every item of equipment used in an electrical installation to comply with the relevant requirements of an applicable British Standard appropriate to the intended use of the equipment. It also offers alternative measures which may be followed where it is not possible to source equipment meeting British or Harmonized Standards, or where it is intended to use an item of equipment in a manner which is not covered by its standard.

133.1

Regulation group 133.2 requires equipment to be suitable in terms of voltage, current, frequency, power, conditions of installation and prevention of harmful effects on other equipment or the supply.

133.2

Part 2   Part 2 contains a number of definitions of relevance to the subject matter of this publication, including:

- consumer unit
- disconnector
- emergency stopping
- emergency switching
- fused connection unit
- isolation
- isolator
- LV switchgear and controlgear assembly
- mechanical maintenance
- monitoring
- switch
- switch, linked
- switch-disconnector

The terms 'isolator' and 'disconnector' are defined as being the same in Part 2 of BS 7671, the definition therein being based upon, but not identical to, the definition of disconnector given in the International Electrotechnical Vocabulary (IEV)

IEC 60050   (IEC 60050).

'Disconnector' is the internationally used term to avoid confusion due to language differences when developing International (IEC) and European (CENELEC) standards. To illustrate this, if one searches for 'insulator' in part 151 (Electrical and magnetic devices / Behaviour and use) of IEC 60050, it can be seen that 'insulator' translates to the following in other languages used by participating countries:

- in French, *isolateur*
- in German, *isolator*
- in Spanish, *aislador*

For convenience, and taking into account traditional usage, the term isolator is used in this Guidance Note.

314.1   Section 314 deals with division of the installation with regard to factors such as
314.3   facilitating the isolation of the installation or parts thereof; permitting safe inspection,
314.2   testing and maintenance; and providing the necessary degree of control to allow the
314.4   installation to be used as intended by the designer.

341.1   Regulation 341.1 requires an assessment to be made of the frequency and quality of maintenance that installed equipment, including switchgear, is likely to receive throughout the life of the installation.

Where an installation location is considered to be difficult to evacuate and/or has a high
422.2.2   density occupation characteristic, Regulation 422.2.2 requires switchgear or controlgear to be only accessible to authorised persons. Where such items are installed within an emergency escape route, an additional enclosure made from non-combustible or not readily combustible material should be provided.

422.3.13   Regulation 422.3.13 requires every circuit within a location having a risk of fire as a result of the nature of processed or stored materials to be provided with a linked switch or linked circuit-breaker to provide isolation from all live conductors, which as defined
537.1.2   includes the neutral conductor. Exceptions to this are given in Regulation 537.1.2,

including circuits forming part of an installation derived from a TN-S or TN-C-S supply; see sections 4.2 and 4.3.

Section 433 deals with provision of devices for protection against overload, and Regulation 433.4 states that if a single protective device protects two or more conductors in parallel then no branch circuits may be installed and no devices providing isolation or switching may be placed in the parallel conductors.

433.4

Amongst other information required to be provided within an installation by Regulation 514.9.1 are those details necessary to allow for the identification of each device provided to perform an isolation or switching function; Regulation 514.11 requires a notice to be posted wherever live parts are present which cannot be isolated by the operation of a single device. Regulation 514.15.1 requires notices to be posted where an installation contains a generator capable of running in parallel with another source.

514.9.1
514.11

514.15.1

As a result of renumbering and reorganisation of the content of the 17th Edition of the Wiring Regulations, the specific requirements relating to isolation and switching are to be found for the most part in Section 537.

Sect 537

The continuity of earthing arrangements is fundamental to many of the measures for protection against electric shock, including the most commonly employed measure, automatic disconnection of supply (ADS). As a result it is not permitted to install a switching device in a protective conductor with the exception of

543.3.4

537.1.5

▶ a changeover arrangement for installations supplied from more than one source of supply which are dependent upon separate earthing arrangements which cannot be connected at the same time
▶ multipole linked switching arrangements and plug-in devices so arranged that the protective conductor
  – cannot be interrupted before the live conductors are disconnected
  – is reconnected before or at the same time as the live conductors.

Regulation 543.4.7 effectively prohibits the installation of an isolator or other switching device in the outer conductor of a concentric cable.

543.4.7

Chapter 55 relates to equipment such as generators, motors and transformers and includes a number of isolation requirements for generators and static convertors.

Chap 55, 551.2.4, 551.4.3.3.3, 551.6.1, 551.7.6

Section 559.10 contains requirements for outdoor lighting installations, highway power supplies and street furniture. Regulation group 559.10.6 gives specific requirements for isolation and switching for such installations.

559.10.6

Where an installation contains safety services, reference needs to be made to Regulation group 560.7 which includes requirements for switchgear and controlgear of such circuits.

560.7

Regarding initial verification, Regulation 611.3(x) requires confirmation of the presence of appropriate isolation and switching devices, suitably placed to be readily operable, as an item on the inspection checklist.

611.3(x)

A number of the sections within Part 7 'Special installations or locations' contain specific requirements for the provision and/or location of isolation and switching devices. These are discussed in detail in Chapter 9 of this Guidance Note.

Part 7

Definitions and principles

3

3.1 General

BS 7671 (the IEE Wiring Regulations) recognises four distinct types of isolation and switching operation:

i isolation
ii switching off for mechanical maintenance
iii emergency switching
iv functional switching.

Each type has specific requirements, and a clear understanding of the meanings of the four types is essential. Devices can fulfil more than one function, but the device selected must fulfil the specific requirements for all the functions needed. Detailed guidance on requirements is given in the following sections, but a brief outline is given below.

Both the Electricity at Work Regulations (EWR) 1989 (Regulation 12(1)(b)) and BS 7671 require that a means of isolation must always be provided for an installation, even in situations where a particular installation may be such that neither switching off for mechanical maintenance nor emergency switching is required. Functional switching is also always necessary.

EWR 1989
132.15.1

The need to provide for isolation, switching off for mechanical maintenance, emergency switching and/or emergency stopping within a particular installation should be considered at an early stage of the design process. The designer will need to consider the intended purpose of the installation, its operational requirements, how it will be used and any specific legislation requirements pertinent to it. Consideration should also be given to the capability of operating and maintenance staff and quality of any maintenance systems in place as regards the devices selected. The designer may, in many cases, be well advised to seek very detailed advice from the owner or user of the installation as to its intended operation and use. See also section 1.2 of this Guidance Note.

132.1

341.1

There is a general requirement that all electrical equipment, including isolating and switching devices, shall be so arranged to facilitate operation, inspection, fault detection, maintenance and repair. Such devices should also be suitable for the environment in which they are expected to be sited and remain operational, and in this regard reference should also be made to Guidance Note 1: *Selection & Erection*.

132.12

512.2

GN1

When positioning isolation and switching devices, the designer should bear in mind the possibility that the designed ease of accessibility may not be maintained when the installation is in use, as it will be beyond the control of the designer once it has been handed over to the client or end user. However, guidance on the use of the installation

© The Institution of Engineering and Technology

should be given in the documentation handed over to the client on completion of the work.

EWR 1989 Regulation 15 of the Electricity at Work Regulations 1989 requires that in order to prevent injury, adequate working space, adequate means of access, and adequate lighting should be provided wherever electrical equipment is sited, so as to permit the safe use of such equipment in situations where danger may arise. Paragraph 7(b) of

HSR25 the guidance given on Regulation 12(3) in HSR25 expands on the need for items of switchgear to remain readily accessible.

511.1 Regulation 511.1 includes the requirement that every item of equipment used in an electrical installation shall comply with the relevant requirements of the applicable British Standard or Harmonized Standard appropriate to the intended use of the equipment.

BS EN 60947 BS EN 60947 *Low-voltage switchgear and controlgear* has multiple parts which cover specific products. These parts contain a number of requirements specific to the isolation function on factors such as performance, indication and marking.

BS EN 60073 BS EN 60073:2002 *Basic and safety principles for man-machine interface, marking and identification. Coding principles for indicators and actuators* provides details and rules for colours, shapes and positions of indicating devices and actuators.

Functional switching devices are mainly covered by:

BS EN 60669-1 ▸ BS EN 60669-1:2004 (equivalent to BS 3676-1:2000) *Switches for household and similar fixed electrical installations: General requirements*

BS EN 60669-2 series ▸ BS EN 60669-2 *Switches for household and similar fixed electrical installations.* Some particular requirements within BS EN 60669 include

BS 5518 -2-1 *Electronic switches* (equivalent to BS 5518:1977 (1999))
-2-2 *Electromagnetic remote-control switches (RCS)*
-2-3 *Time delay switches (TDS)*
-2-4 *Isolating switches*

BS 1363 ▸ BS 1363:1995 *13 A plugs, socket-outlets, adaptors and connection units*
BS EN 60309-2 ▸ BS EN 60309-2:1999 *Plugs, socket-outlets and couplers for industrial purposes. Dimensional interchangeability requirements for pin and contact-tube accessories*
BS 5733 ▸ BS 5733:1995 *Specification for general requirements for electrical accessories.*

However, there are several systems, such as electronic lighting control by building management systems, which may be made up from many units or components that have no specific product standard. In such cases, the individual units or components should be manufactured and installed to relevant standards and the installation should comply with BS 7671. For home and building electronic systems (HBES) see also

BS EN 50428 BS EN 50428:2005 +A1:2007.

511.2 Where equipment to be used is not covered by a British Standard or Harmonized Standard or is used outside the scope of its standard, the designer or other person responsible for specifying the installation must confirm that the equipment provides at least the same degree of safety as that afforded by compliance with the Wiring Regulations. The designer or other person responsible may be liable to prosecution if an accident is found to have been caused by inappropriately selected equipment.

Specific legislation must also be complied with. This may include specific duties or requirements and overrides any other requirements or standards.

Table 53.2, newly introduced in the 17th Edition, gives guidance on the selection of appropriate devices to perform protective, isolation and switching functions and is reproduced here as Table 3.1.

Table 53.2

▼ **Table 3.1**
Guidance on the selection of protective, isolation and switching devices

Device	Standard	Isolation[5]	Emergency switching [2]	Functional switching
Switching device	BS 3676: Pt 1 1989	Yes[4]	Yes	Yes
	BS EN 60669-1	No	Yes	Yes
	BS EN 60669-2-1	No	No	Yes
	BS EN 60669-2-2	No	Yes	Yes
	BS EN 60669-2-3	No	Yes	Yes
	BS EN 60669-2-4	Yes	Yes	Yes
	BS EN 60947-3	Yes[1]	Yes	Yes
	BS EN 60947-5-1	No	Yes	Yes
Contactor	BS EN 60947-4-1	Yes[1]	Yes	Yes
	BS EN 61095	No	No	Yes
Circuit-breaker	BS EN 60898	Yes	Yes	Yes
	BS EN 60947-2	Yes[1]	Yes	Yes
	BS EN 61009-1	Yes	Yes	Yes
RCD	BS EN 60947-2	Yes[1]	Yes	Yes
	BS EN 61008-1	Yes	Yes	Yes
	BS EN 61009-1	Yes	Yes	Yes
Isolating switch	BS EN 60669-2-4	Yes	Yes	Yes
	BS EN 60947-3	Yes	Yes	Yes
Plug and socket-outlet (≤ 32 A)	BS EN 60309	Yes	No	Yes
	IEC 60884	Yes	No	Yes
	IEC 60906	Yes	No	Yes
Plug and socket-outlet (> 32 A)	BS EN 60309	Yes	No	No
Device for the connection of luminaire	BS IEC 61995-1	Yes[3]	No	No
Control and protective switching device for equipment (CPS)	BS EN 60947-6-1	Yes	Yes	Yes
	BS EN 60947-6-2	Yes[1]	Yes	Yes
Fuse	BS 88	Yes	No	No
Device with semiconductors	BS EN 60669-2-1	No	No	Yes
Luminaire Supporting Coupler	BS 6972	Yes[3]	No	No
Plug and unswitched socket-outlet	BS 1363-1	Yes[3]	No	Yes
	BS 1363-2	Yes[3]	No	Yes
Plug and switched socket-outlet	BS 1363-1	Yes[3]	No	Yes
	BS 1363-2	Yes[3]	No	Yes
Plug and socket-outlet	BS 5733	Yes[3]	No	Yes
Switched fused connection unit	BS 1363-4	Yes[3]	Yes	Yes
Unswitched fused connection unit	BS 1363-4	Yes[3] (Removal of fuse link)	No	No
Fuse	BS 1362	Yes	No	No
Cooker control unit switch	BS 4177	Yes	Yes	Yes

Yes	Function provided
No	Function not provided

[1] Function provided if the device is suitable and marked with the symbol for isolation (see BS EN 60617 identity number S00288). ⟋⊢

[2] See Regulation 537.4.2.5.

[3] Device is suitable for on-load isolation, i.e. disconnection whilst carrying load current.

[4] Function provided if the device is suitable and marked with ◎.

[5] In an installation forming part of a TT or IT system, isolation requires disconnection of all the live conductors. See Regulation 537.2.2.1.

3.2   Isolation

The definition given in BS 7671 is:

Part 2   *A function intended to cut off for reasons of safety the supply from all, or a discrete section, of the installation by separating the installation or section from every source of electrical energy.*

EWR 1989   Regulation 12(1) of the Electricity at Work Regulations 1989 requires that:

> (1) Where necessary to prevent danger, suitable means (including, where appropriate, methods of identifying circuits) shall be available for –
> (a) cutting off the supply of electrical energy to any electrical equipment; and
> (b) the isolation of any electrical equipment.

Regulation 12(2) defines isolation as:

> *… the disconnection and separation of the electrical equipment from every source of electrical energy in such a way that this disconnection and separation is secure.*

HSR25   The *Memorandum of guidance on the Electricity at Work Regulations 1989* (HSR25) published by the Health and Safety Executive advises, with reference to Regulation 12(1)(b) above, that isolation is the process of ensuring that the supply to all or a particular part of an installation remains switched off and that inadvertent reconnection is prevented.

The issue of preventing inadvertent reconnection is covered in Regulation 13:

> **13. Precautions for work on equipment made dead**
>
> Adequate precautions shall be taken to prevent electrical equipment, which has been made dead in order to prevent danger while work is being carried out on or near that equipment, from becoming electrically charged during that work if danger may thereby arise.

The 'security' mentioned in Regulation 12(2) may be achieved either as a result of the means of isolation remaining directly under the control of the persons who are reliant upon its remaining effective or, where this is not the case, by the application of a locking device being used to secure the means of isolation in the OFF position.

Neither the Electricity at Work Regulations 1989 nor BS 7671 calls for the application of a means of locking to be applied in all cases. However, if any possibility exists of the means of isolation being compromised, some form of locking should be applied.

537.2.1.5   Indeed, Regulation 537.2.1.5 requires the means of isolation to be locked-off where it is placed remotely from the equipment being isolated by its operation.

132.15.1   The statutory requirements given in Regulations 12 and 13 of the EWR 1989 are broadly repeated in Regulation 132.15.1 of BS 7671.

The complexity of isolation and switching arrangements increases with the size of the electrical installation. Figure 3.1 shows switchboards suitable for use in large industrial or commercial installations.

▼ **Figure 3.1**
Switchboards suitable for use in industrial or commercial installations awaiting delivery to site [photograph courtesy of Schneider]

3.3 Switching off for mechanical maintenance

The term 'mechanical maintenance' is defined in BS 7671 as:

> *The replacement, refurbishment or cleaning of lamps and non-electrical parts of equipment, plant and machinery.* Part 2

Broadly speaking, switching off for mechanical maintenance is a function similar to isolation whereby electrically actuated equipment is made safe for persons to work on, in, or near the equipment, whether or not they are electrically skilled.

Although there is no reference to this operation in the Electricity at Work Regulations, the guidance given on Regulation 12(1)(a) in the *Memorandum of guidance on the Electricity at Work Regulations 1989* (HSR25) acknowledges that 'there may be a HSR25
need to switch off electrical equipment for reasons other than preventing electrical danger but these considerations are outside the scope of the [Electricity at Work] Regulations'.

It should be remembered at this point that the scope of BS 7671 includes the 131.1(v)
protection of persons from risk of injury from mechanical movement of electrically actuated equipment, by the employment of electrical emergency switching or electrical switching for mechanical maintenance of non-electrical parts of such equipment.

The purpose of the measure is to prevent physical injury, but not electric shock or burns, as mechanical maintenance should not involve work upon, access to, or exposure of normally live parts. Particular attention is drawn therefore to the requirements of Regulations 14 and 16 of the EWR 1989 in respect of working on or near live parts.

Switching off for mechanical maintenance should be considered if access to machinery or equipment may involve access to normally moving parts. Isolation of supplies to machinery or equipment may be more appropriate in some situations to provide a sufficient degree of physical safety.

3.4   Emergency switching

BS 7671 defines this as:

Part 2

An operation intended to remove, as quickly as possible, danger, which may have occurred unexpectedly.

537.4.1 It is recognised that this danger may arise from non-electrical events or occurrences and it is for this reason that Regulation 537.4.1, which will be considered later in this publication, is accompanied by a note stating that 'Emergency switching may be emergency switching on or emergency switching off'.

3.5   Emergency stopping

The definition given in BS 7671 is:

Part 2

Emergency switching intended to stop an operation.

537.4.2.2 Note 2
HSR25 Again, it should be appreciated that the operation of an emergency stopping device may, in some cases, need to allow the continued supply to electrically actuated brakes and as such not remove ALL sources of supply. This is obviously an important factor when considering suitable means for achieving safe isolation. HSR25 reminds readers of this in its discussion of Regulation 12(1)(b), where it states that 'it must be understood that the two functions of switching off (which includes emergency stopping) and isolation are not the same, even though in some circumstances they are performed by the same action or by the same equipment'.

PUWER Regulation 16 of the Provision and Use of Work Equipment Regulations 1998 provides specific requirements for machine emergency stop controls and reference should also be made to the following standards:

BS EN ISO 13850 ▶ BS EN ISO 13850:2008 *Safety of machinery. Emergency stop equipment, functional aspects. Principles for design*

BS EN 60204-1 ▶ BS EN 60204-1:2006 (2009) *Safety of machinery. Electrical equipment of machines. General requirements*

BS EN 60947-5-5 ▶ BS EN 60947-5-5:1998 *Low-voltage switchgear and controlgear. Control circuit devices and switching elements. Electrical emergency stop devices with mechanical latching function.*

3.6   Functional switching

This is defined in BS 7671 as:

Part 2

An operation intended to switch ON or OFF or vary the supply of electrical energy to all or part of an installation for normal operating purposes.

537.5 As BS 7671 is mainly concerned with the safety of electrical installations it only deals with the operation of functional switching and functional switching devices generally.

BS 7671 does, however, recognise that some functional switching or control devices may also be employed to provide variously the functions of isolation, emergency switching, or switching off for mechanical maintenance purposes. However, it is incumbent upon the designer to confirm that any device selected to provide more than one switching function in an installation meets the relevant requirements of BS 7671 for each such function. Table 3.1 (Table 53.2 of BS 7671) provides guidance on the selection of protective, isolation and switching devices.

Table 53.2

3.7 Ordinary, instructed and skilled persons

These are defined in BS 7671.

Part 2

Ordinary person. *A person who is neither a skilled person nor an instructed person.* This would include householders and other persons who are neither instructed nor skilled persons (as defined).

Instructed person. *A person adequately advised or supervised by skilled persons to enable him/her to avoid dangers which electricity may create.*

Skilled person. *A person with technical knowledge or sufficient experience to enable him/her to avoid dangers which electricity may create.*

3.8 Definitions from other standards

Table 53.2 lists the product standards appropriate to the commonly available isolation and switching devices. These standards contain further definitions taken directly from, or based upon, those given in the International Electrotechnical Vocabulary (IEV) (IEC 60050), a number of which are reproduced below.

Table 53.2

IEC 60050

i   Breaking capacity (of a switching device or fuse)
Value of prospective breaking current that a switching device or a fuse is capable of breaking at a stated voltage under prescribed conditions of use and behaviour.

Note 1: The voltage to be stated and the conditions to be prescribed are dealt with in the relevant product standard.
Note 2: For a.c., the current is expressed as the symmetrical rms value of the a.c. component.

For *short-circuit breaking capacity*, see **viii** below.

ii   Breaking current (of a switching device or fuse)
The current in a pole of a switching device or in a fuse at the instant of initiation of the arc during a breaking process.

Note: For a.c., the current is expressed as the symmetrical rms value of the a.c. component.

iii   Conditional short-circuit current (of a circuit or switching device)
Prospective current that a circuit or a switching device, protected by a specified short-circuit protective device, can satisfactorily withstand for the total operating time of that device under specified conditions of use and behaviour.

Note 1: For the purpose of this standard, the short-circuit protective device is generally a circuit-breaker or a fuse.
Note 2: This definition differs from IEV 441-17-20 by broadening the concept of current-limiting device into a short-circuit protective device, the function of which is not only to limit the current.

iv Fuse-combination unit
Combination of a mechanical switching device and one or more fuses in a composite unit, assembled by the manufacturer or in accordance with his instructions.

v Making capacity (of a switching device)
Value of prospective making current that a switching device is capable of making at a stated voltage under prescribed conditions of use and behaviour.

Note: The voltage to be stated and the conditions to be prescribed are dealt with in the relevant specifications.

vi Rated value
A quantity value assigned, generally by the manufacturer, for a specified operating condition of a component, device or equipment.

Note: Examples of rated value usually stated for fuses: voltage, current, breaking capacity.

vii Rating
The set of rated values and operating conditions.

viii Short-circuit breaking capacity
Breaking capacity for which prescribed conditions include a short-circuit at the terminals of the switching device.

ix Short-circuit making capacity
Making capacity for which the prescribed conditions include a short-circuit at the terminals of the switching device.

x Switch-disconnector
Switch which, in the open position, satisfies the isolating requirements specified for a disconnector.

xi Utilisation category (for a switching device or a fuse)
A combination of specified requirements related to the condition in which the switching device or the fuse fulfils its purpose, selected to represent a characteristic group of practical applications.

Note: The specified requirements may concern, for example, the values of making capacities (if applicable), breaking capacities and other characteristics, the associated circuits and the relevant conditions of use and behaviour.

The utilisation category of equipment defines the intended application and is specified in the relevant product standard and is characterised by one or more of the following service conditions:

- ▶ current(s), expressed as multiple(s) of the rated operational current
- ▶ voltage(s), expressed as multiple(s) of the rated operational voltage
- ▶ power factor or time constant
- ▶ short-circuit performance
- ▶ selectivity
- ▶ other service conditions, as applicable.

BS EN 60947-1   Annex A of BS EN 60947-1:2007 gives examples of utilisation categories for low voltage switchgear and controlgear and is reproduced here as Table 3.2.

▼ **Table 3.2**　Examples of utilisation categories for low voltage switchgear and controlgear

Nature of current	Category	Typical applications	Relevant IEC product standard
a.c.	AC-20	Connecting and disconnecting under no-load conditions	
	AC-21	Switching of resistive loads, including moderate overloads	
	AC-22	Switching of mixed resistive and inductive loads, including moderate overloads	60947-3
	AC-23	Switching of motor loads or other highly inductive loads	
a.c.	AC-1	Non-inductive or slightly inductive loads, resistance furnaces	
	AC-2	Slip-ring motors: starting, switching off	
	AC-3	Squirrel-cage motors: starting, switching off motors during running	
	AC-4	Squirrel-cage motors: starting, plugging,[a] inching[b]	
	AC-5a	Switching of electric discharge lamp controls	
	AC-5b	Switching of incandescent lamps	60947-4-1
	AC-6a	Switching of transformers	
	AC-6b	Switching of capacitor banks	
	AC-8a	Hermetic refrigerant compressor motor control with manual resetting of overload releases	
	AC-8b	Hermetic refrigerant compressor motor control with automatic resetting of overload releases	
a.c.	AC-52a	Control of slip-ring motor stators: 8 h duty with on-load currents for start, acceleration, run	
	AC-52b	Control of slip-ring motor stators: intermittent duty	
	AC-53a	Control of squirrel-cage motors: 8 h duty with on-load currents for start, acceleration, run	
	AC-53b	Control of squirrel-cage motors: intermittent duty	60947-4-2
	AC-58a	Control of hermetic refrigerant compressor motors with automatic resetting of overload releases: 8 h duty with on-load currents for start, acceleration, run	
	AC-58b	Control of hermetic refrigerant compressor motors with automatic resetting of overload releases: intermittent duty	
a.c.	AC-51	Non-inductive or slightly inductive loads, resistance furnaces	
	AC-55a	Switching of electric discharge lamp controls	
	AC-55b	Switching of incandescent lamps	60947-4-3
	AC-56a	Switching of transformers	
	AC-56b	Switching of capacitor banks	
a.c.	AC-12	Control of resistive loads and solid-state loads with isolation by optocouplers	
	AC-13	Control of solid-state loads with transformer isolation	
	AC-14	Control of small electromagnetic loads	60947-5-1
	AC-15	Control of a.c. electromagnetic loads	
a.c.	AC-12	Control of resistive loads and solid-state loads with optical isolation	
	AC-140	Control of small electromagnetic loads with holding (closed) current ≤ 0.2 A, e.g. contactor relays	60947-5-2
a.c.	AC-31	Non-inductive or slightly inductive loads	
	AC-33	Motor loads or mixed loads including motors, resistive loads and up to 30% incandescent lamp loads	
	AC-35	Electric discharge lamp loads	60947-6-1
	AC-36	Incandescent lamp loads	

▼ **Table 3.2**   *continued*

Nature of current	Category	Typical applications	Relevant IEC product standard
a.c.	AC-40	Distribution circuits comprising mixed resistive and reactive loads having a resultant inductive reactance	60947-6-2
	AC-41	Non-inductive or slightly inductive loads, resistance furnaces	
	AC-42	Slip-ring motors; starting, switching off	
	AC-43	Squirrel-cage motors: starting, switching off motors during running	
	AC-44	Squirrel-cage motors: starting, plugging,[a] inching[b]	
	AC-45a	Switching of electric discharge lamp controls	
	AC-45b	Switching of incandescent lamps	
a.c.	AC-7a	Slightly inductive loads for household appliances and similar applications	61095
	AC-7b	Motor loads for household applications	
a.c. and d.c.	A	Protection of circuits, with no rated short-time withstand current	60947-2
	B	Protection of circuits, with a rated short-time withstand current	
d.c.	DC-20	Connecting and disconnecting under no-load conditions	60947-3
	DC-21	Switching of resistive loads, including moderate overloads	
	DC-22	Switching of mixed resistive and inductive loads, including moderate overloads (e.g. shunt motors)	
	DC-23	Switching of highly inductive loads (e.g. series motors)	
d.c.	DC-1	Non-inductive or slightly inductive loads, resistance furnaces	60947-4-1
	DC-3	Shunt-motors, starting, plugging,[a] inching.[b] Dynamic braking of motors	
	DC-5	Series-motors, starting, plugging,[a] inching.[b] Dynamic braking of motors	
	DC-6	Switching of incandescent lamps	
d.c.	DC-12	Control of resistive loads and solid-state loads with isolation by optocouplers	60947-5-1
	DC-13	Control of electromagnets	
	DC-14	Control of electromagnetic loads having economy resistors in circuit	
d.c.	DC-12	Control of resistive loads and solid-state loads with optical isolation	60947-5-2
	DC-13	Control of electromagnets	
d.c.	DC-31	Resistive loads	60947-6-1
	DC-33	Motor loads or mixed loads including motors	
	DC-36	Incandescent lamp loads	
d.c.	DC-40	Distribution circuits comprising mixed resistive and reactive loads having a resultant inductive reactance	60947-6-2
	DC-41	Non-inductive or slightly inductive loads, resistance furnaces	
	DC-43	Shunt-motors: starting, plugging,[a] inching.[b] Dynamic braking of d.c. motors	
	DC-45	Series-motors: starting, plugging,[a] inching.[b] Dynamic braking of d.c. motors	
	DC-46	Switching of incandescent lamps	

Notes:

a By 'plugging' is understood stopping or reversing the motor rapidly by reversing motor primary connections while the motor is running.

b By 'inching' (jogging) is understood energizing a motor once or repeatedly for short periods to obtain small movements of the driven mechanism.

Isolating switches complying with BS EN 60669-2-4 are intended for use in household and similar fixed electrical installations. BS EN 60669-2-4

Isolating switches must be marked with the symbols for isolating function and for the closed and open positions.

  Symbol for isolating function to BS EN 60669-2-4

The isolating switch must be tested to verify that it is able to withstand, without damage, short-circuit currents up to and including its rated conditional short-circuit current (I_{nc}).

The marking for the rated conditional short-circuit current (I_{nc}) must be on the isolating switch or in the manufacturer's published literature.

The manufacturer must provide reference(s) of one or more short-circuit protection devices (SCPDs) in their catalogue and/or in the instructions which are provided with the isolating switch.

An isolating switch to BS EN 60669-2-4 must be subjected to minimum values of energy let-through (I^2t) and peak current (I_p), as stipulated in Table 102 of the standard. Similarly, Table 17 of BS EN 60669-2-4 specifies the minimum number of on-load operations in relation to the rated current.

3.9 General requirements for warning notices and labelling

Adequate notices, labels and instructions must be provided to ensure the safe and proper operation of the electrical installation, both under normal operating conditions and those likely in an emergency. There is a general requirement that, except for instances where no possibility of confusion exists, a label or other suitable means of identification should be installed to indicate the purpose of each item of switchgear or controlgear. BS 7671 also contains a number of more specific requirements which are considered below in the relevant sections. 132.13 514.1.1 514.11.1

Although not specifically covered by BS 7671, there is also a need for temporary warning labels to be posted to inform people when equipment has been shut down, such as when circuits have been isolated or when particular maintenance tasks are being carried out. Any such signs must be clearly and prominently displayed. An example of such a warning label is given in Appendix B of this publication.

Labels and warning notices should be of the size required and of a type suitable for the location. They should be installed in such a way that they are unlikely to be painted over or easily removed or defaced. In some cases, labels and notices should be permanently fixed by suitable screws or rivets, taking care not to damage equipment, invalidate IP ratings or block vents. Self-adhesive labels may be used. However, care should be taken to ensure that they remain permanently attached and are not adversely affected by factors such as ambient temperature, sunlight and humidity. Section 514 537.6.4 705.537.2 712.537.2.2.5.1

Any switchgear and controlgear provided in connection with safety services should be clearly identified. 560.7.5

Isolation

4

4.1 General

The fundamental principle for isolation is that effective means suitably placed for ready operation shall be provided so that all voltage may be cut off from every installation, from every circuit thereof and from all equipment, as may be necessary to prevent or remove danger. As a result of the restructuring within the 17th Edition, the majority of requirements relating to isolation are now to be found in Section 537. **131.15.1** **Section 537**

In the case of locations with risks of fire due to the nature of processed or stored materials, such as woodworking shops, industrial scale bakeries, paper mills and the like, every circuit, subject to the exceptions given in Regulation 537.1.2, should be provided with a means of isolation which disconnects all live conductors by means of a linked switch or linked circuit-breaker **422.3.13** **537.1.2**

Fuses and associated neutral links may be used to facilitate isolation. However, the use of fuses and links should be restricted to skilled or instructed persons (as defined) to avoid any risks associated with their being withdrawn or replaced on load, and the following measures should be observed in order to minimise the risk of unauthorised and/or premature reconnection:

▶ Distribution boards from which protective devices are removed for the purpose of isolation must be locked or otherwise secured, as they will usually be sited remote from the equipment being isolated. This should normally be achieved by a locking mechanism on the distribution board itself. In the rare event of the distribution board not having a locking facility (typically because of the age of the board) then the room containing the distribution board should be locked. In either case, the key should remain with the person relying upon the isolation remaining effective at all times
▶ Any fuses and links removed for the purpose of isolation should not be left in the vicinity of the distribution board and should remain under the control of the person relying upon the isolation remaining effective in a similar way to that described for a key securing the means of isolation. Fuseway blanking pieces (preferably lockable) are available and may be utilised for warning.

In general, whichever form of isolation is used (fuse withdrawal, circuit-breaker operation, etc.) it is good practice to apply a caution notice at the point of isolation.

A particular danger may arise in older single-phase installations where double-pole fuseboards may still be found. To avoid any confusion it is recommended that both fuses protecting a circuit are removed. In any case, isolation should be proved by the use of an approved voltage detector. Double-pole fuseboards were common in single-phase and neutral installations which would now be of a considerable age, greater than the expected life of such installations, and a re-wire would be advisable for such installations where they are found. The owner of the installation should be informed and matters noted on any Certificate or Report issued. Double-pole fuseboards may also be found in certain specialist installations.

514.11.1 Where an installation, item of equipment or enclosure contains live parts which require the operation of more than one isolating device in order to disconnect all sources of supply, a notice warning of the need to disconnect more than one supply should be posted such that it will be seen by a person wishing to gain access to said installation,

537.2.1.3 item of equipment or enclosure. See Figure 4.1.

▼ **Figure 4.1**
Examples of labelling where one switch does not isolate equipment

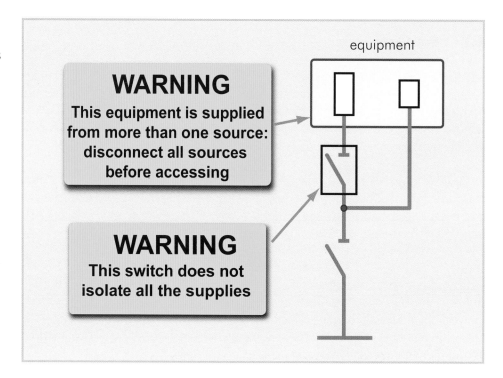

4.2   Isolation at the origin of an installation

537.1.3 Each installation must be provided with a means of disconnecting from the supply.

537.1.4 This means of isolation, either a linked switch or linked circuit breaker, should be placed as near as practicable to the origin of the installation and should be suitable to allow for switching the supply on load. In single-phase domestic and similar properties to be used by ordinary persons (as defined in BS 7671) this switchgear should be double-pole so as to break both the live conductors. See Figures 4.2 and 4.3.

537.1.2 In general, the neutral conductor of a supply derived from a TN-S or TN-C-S system in accordance with the Electricity Safety, Quality and Continuity Regulations 2002 can be considered to be reliably connected to Earth by a suitably low impedance. Where this is the case, it is not necessary to isolate or switch the neutral conductor except as mentioned above – see Figures 4.2 and 4.5. Combined protective and neutral (PEN) conductors are not commonplace in installations within the UK. Where they are installed, they should not be isolated or switched.

a

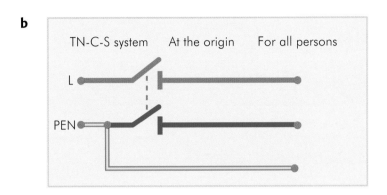

b

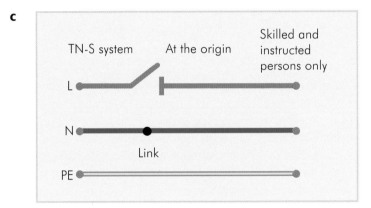

c

d

▼ **Figure 4.2**
Isolation at the origin
of typical single-phase
installations

537.1.4

537.1.2

▼ **Figure 4.3**
Isolation at the origin
of an installation within
a single-phase TT or IT
system

537.1.2

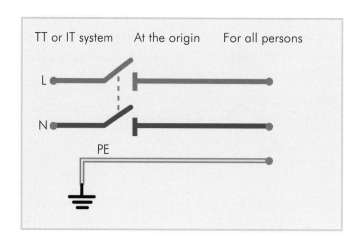

▼ **Figure 4.4**
Isolation at the origin of
an installation supplied
from a d.c. source
– all poles are to be
provided with a means
of isolation

537.1.2

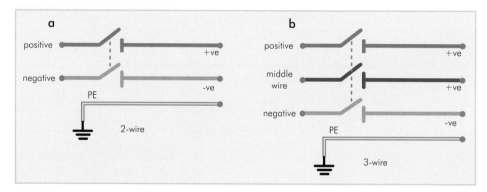

537.1.5 Where an installation is supplied from more than one source, which is increasingly the case even for domestic and other small premises, and these sources are required to have their own independent earthing arrangements, a changeover switching arrangement should be provided between the neutral point and means of earthing to ensure that not more than one means of earthing is effective at any one time. This changeover of the earth connection should occur at substantially the same time as the changeover of the associated live conductors.

537.1.6 A main switch is required for each source of supply to an installation. These main switches may be arranged such that they can be operated at the same time, via a suitable interlock arrangement, to provide disconnection of the installation from all sources of supply. Where this is not the case, a warning notice informing of the need to operate all main switches to fully isolate the installation will be required.

It is permitted for a dwelling to have more than one electrical installation. Therefore one 'main switch' is not required to isolate all consumer units simultaneously provided the consumer units have an integral main switch.

537.2.2.6 This situation frequently occurs when an additional consumer unit is added, e.g. to supply an electric shower. Regulation 537.2.2.6 requires each device used for isolation to be clearly identified by position or durable marking to indicate the installation it isolates.

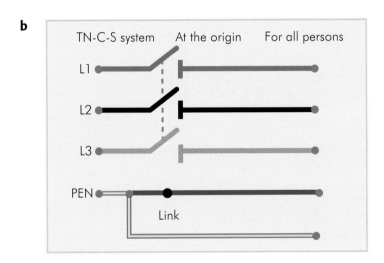

a TN-S system At the origin For all persons

L1

L2

L3

N Link

PE

b TN-C-S system At the origin For all persons

L1

L2

L3

PEN Link

c TT or IT system At the origin For all persons

L1

L2

L3

N

PE

▼ **Figure 4.5**
Isolation at the origin
of a three-phase
installation

537.1.4

537.1.4

4.3 Isolation of circuits within an installation

537.2.1.1 Provision should be made such that each circuit within an installation can be isolated from its source(s) of supply. As was the case with reference to Regulation 537.1.2 (discussed above) for main isolation, in an installation supplied from a TN-S or TN-C-S system, this will not require the neutral to be switched. See Figures 4.6(a) and (c).

It is permitted to isolate more than one circuit by a single device. This might be appropriate, for example, in the case of a production line consisting of a number of pieces of equipment controlled from a single control panel.

537.2.1.2 Regulation 537.2.1.2 requires suitable means to be provided to prevent any item of equipment from becoming inadvertently or unintentionally energized. It is generally assumed that this will be achieved by either locking the means of isolation in the OFF position, or on account of the device remaining under the direct control of the persons reliant upon the isolation remaining effective.

537.2.1.4 On occasion, equipment may contain components which retain a charge for some time after the supply has been disconnected. Where this is the case, a means should be provided to effect their discharge.

537.2.1.5 It is quite acceptable for an isolator to be placed remotely from the equipment it isolates. However, where this is the case, the means of isolation must be capable of being secured in the OFF or open position.

This security may be provided by:

▶ the operation and removal of a unique, non-interchangeable key or handle (see Figures 4.9 and 4.11)
▶ a padlock or similar
▶ the removal of one or more fuses and any 'spare' fuses in the immediate vicinity
▶ locking the lid of a distribution board
▶ locking the door to a switchroom (where the installation is under the control of skilled persons and an effective system of work or permit-to-work is in place – see Appendix C).

537.2.1.7
537.2.2.4 Although it is not a requirement to isolate or switch the neutral conductor in installations supplied from a TN supply system, it is still necessary to provide a means for disconnecting the neutral conductor. This requirement is normally met by the provision of a terminal or removable link which must remain accessible to skilled persons throughout the life of the installation. It should only be possible to disconnect the neutral conductor by the use of a tool, *not* a coin or similar.

560.7.5 Switchgear and controlgear provided in connection with a safety service should be installed in a plant room or similar such that it is accessible only to skilled or instructed persons (as defined).

BS 5839-1
BS 5839-6 Reference should also be made to the relevant standard for the safety service in question. In the case of fire alarm systems in industrial and commercial premises, reference should also be made to the requirements of clause 25.2 of BS 5839-1:2002 (2008) and for fire detection and alarm systems in domestic premises, section 15 of BS 5839-6:2004.

Single-phase

a

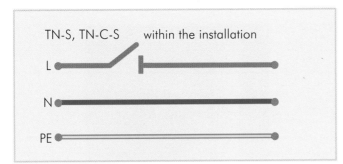

b

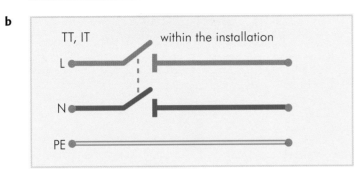

Three-phase

c

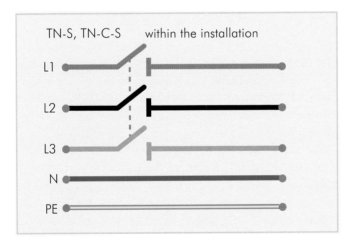

d

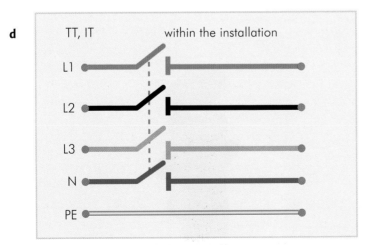

▼ **Figure 4.6**
Examples showing isolation arrangements for circuits within an installation

537.2.1.1

537.2.1.1

In the case of emergency lighting installations, reference should be made to clause 9.3 of BS 5266-1:2005.

BS 5266-1

4.4   Devices for isolation and securing isolation

▼ Figure 4.7
Switch-disconnector
capable of being locked
in the OFF position
[photograph courtesy
of Hager]

▼ Figure 4.8
Three types of lock-out
designed for use with
a padlock to secure
circuit-breakers, RCBOs
and RCDS

537.1.2

537.2.2.1

Apart from installations supplied from a TN supply system in accordance with the Electricity Safety, Quality and Continuity Regulations 2002 where disconnection of the neutral conductor is not required, Regulation 537.2.2.1 requires the provision of a device to isolate all live supply conductors from the circuit concerned. See Figures 4.6(b) and (d).

The device so selected must be suitable for the overvoltage category appropriate at its point of installation.

This regulation also makes it clear that semiconductor devices may not be used as isolating devices as they do not provide any physical break in the conductor let alone one of the dimensions required for an isolating device. They may, however, be used to initiate the operation of an isolator.

▼ Figure 4.9
Removable key switch
type isolator suitable for
off-load operation only
and for use by skilled or
instructed persons (as
defined) [photograph
courtesy of Castell]

Isolators are provided to put a deliberate break in conductors in order to disconnect the supply to equipment or accessories downstream. If the position of the contacts or other means of providing this break in continuity is not externally visible (which is not common in modern switchgear), a clear and reliable indication of their position, which occurs only once the isolated position has been achieved, must be provided.

537.2.2.2

A device selected to act as an isolator should be either designed or installed such that unintentional or inadvertent closure is prevented.

537.2.2.3

▼ **Figure 4.10**
Lock-out designed to secure ordinary rocker-type switches, such as those found on some older small switchgear and consumer units

Isolators which are only suitable for use off-load should be capable of being secured in the OFF position to prevent inadvertent or unauthorised operation by persons who are not electrically skilled.

537.2.2.4

Although it is preferred that the means of isolation is a multipole switching device which, when operated, disconnects all applicable poles of the supply, the use of single-pole devices placed adjacent to each other is also permitted.

537.2.2.5

▼ **Figure 4.11**
Isolator of the removable key type, suitable for on-load operation and for use by ordinary persons (as defined) [photograph courtesy of Castell]

The purpose of every device that is to be used to provide isolation should be clear either as a direct result of its position, as would be the case with an isolator forming an integral part of an item of equipment, or as a result of labelling, which would be appropriate where an isolator was mounted in a location remote from the item of equipment which it controls. In such a case it would, in all but the most simple of installations, be necessary to place a notice local to both the isolator and the item of equipment in question.

537.2.2.6

▼ **Figure 4.12**
Illustration of switched fused connection unit secured by locking the fuse carrier

Table 53.2　A plug and socket-outlet may be used to provide isolation of an individual item of equipment.

▼ **Figure 4.13**
Lock-outs with the facility for more than one lock to be applied at one time

4.5　Isolation requirements for particular types of equipment

4.5.1　High voltage discharge lighting

537.2.1.6　Where an installation contains circuits supplying high voltage (as defined in BS 7671) discharge lighting, one or more of the following means of isolation should be provided:

▶ an interlock system arranged such that the supply is disconnected before access can be gained to live parts. The operation of the switch may be initiated, for example, by the action of removing an access panel or opening a door and is additional to any control switching device provided
▶ a local isolator, again additional to any control switching device provided
▶ a switch having a removable handle or facility for locking; or by securing the access door of a distribution board using a lock or removable handle. The means of locking should meet the non-interchangeability requirements of Regulation 537.2.1.5 discussed in section 4.3.

The requirements relating to the installation of firefighters' switches in connection with high voltage discharge lighting circuits are discussed in Chapter 8 of this publication.

4.5.2 Generators

Where an installation contains its own generation capability the requirements of Section 551 must be taken into consideration, in particular those of Regulation 551.6.1 regarding measures to prevent a generator designed to operate as a switched alternative (standby) source of supply from being operated in parallel with a supply from a public supply system. This may be achieved for example by:

Section 551
551.6.1

▶ an interlock arrangement between the operating mechanisms or control circuits of the changeover switches – the interlock may be electrical, mechanical or electromechanical in operation
▶ a system of locks having a single transferable key
▶ a three position break-before-make changeover switch
▶ an automatic changeover switch and interlock.

Where an installation contains a generator arranged to provide an additional source of supply in parallel with another source it is necessary to post the warning notice shown in Figure 4.14 in the following locations within the installation to raise awareness of the need to isolate more than one source of supply:

514.15.1

▶ at the origin
▶ at the meter position, if this is remote from the origin
▶ at the distribution board or consumer unit to which the generator is connected
▶ at ALL points of isolation provided for BOTH sources of supply.

▼ **Figure 4.14**
Example of dual supply warning label (Regulation 514.15.1)

It is becoming increasingly common for small-scale generators to be installed within even small installations such as domestic premises.

Where this is the case, it is necessary to provide a means of:

▶ automatic switching that will disconnect the generator from the public electricity distribution system in the event of the public supply failing or exceeding the acceptable variation of voltage and frequency given in Table 1 of Energy Networks Association (ENA) G83/1-1, reproduced as Table 4.1
▶ isolating the generator from the public distribution network (see Figure 4.15). For generators of output exceeding 16 A, this means of isolation must comply with relevant national rules and the requirements of distribution network operators (DNOs), currently, at the time of publication of this Guidance Note, ETR 113 and G59/1 – both published by the ENA.

551.7.4

G83/1-1
551.7.6

ETR 113
G59/1

▼ **Table 4.1**
Protection settings
[table courtesy of ENA]

Parameter	Trip setting (maximum range)	Trip time (maximum value)*
Over voltage	264 volts (230 +14.7%)	1.5 seconds
Under voltage	207 volts (230 −10%)	1.5 seconds
Over frequency	50.5 Hz (50 Hz +1%)	0.5 second
Under frequency	47 Hz (50 Hz −6%)	0.5 second
Loss of mains	Note 3	0.5 second

Notes:

* For each protection function listed in Table 4.1 it is permissible to extend the relay operating time to 5 seconds for those SSEG units that can withstand being re-energized from a source that is 180° out of phase with the SSEG output. Typically this will only be applicable to SSEG units connected via an invertor e.g. a PV array.

1 Voltage and frequency is referenced to the supply terminals.

2 To reduce the risk of nuisance tripping for normal system disturbances of short duration, the trip times for under and over voltage have been extended from the 0.5 second specified in Engineering Recommendation G59/1. In the event of the SSEG being installed on a circuit controlled by an auto-reclosing circuit-breaker, the Loss of Mains protection will ensure that the SSEG is disconnected before the circuit-breaker can reclose after tripping on fault.

3 The loss of mains protection shall use a recognised technique (as defined in the relevant Annex of G83/1-1). Active methods which use impedance measuring techniques by drawing current pulses from, or injecting a.c. currents into, the DNO's system are not considered to be suitable.

▼ **Figure 4.15**
Isolation arrangements
for small-scale generator

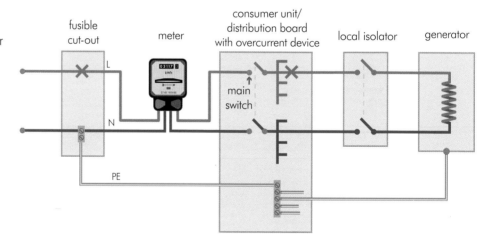

BS EN 50438

For generators of output not exceeding 16 A, the accessibility requirements of clause 4.2.1.3 of BS EN 50438:2007 *Requirements for the connection of micro-generators in parallel with public low-voltage distribution networks* applies. This states that where the means of isolation is not accessible to the DNO at all times, it is acceptable to provide two means of automatic disconnection having a single control. At least one of these means of disconnection must be afforded by the physical separation of mechanical contacts.

4.5.3   Motors

A fundamental requirement is that every fixed electric motor should be provided with an efficient means of switching off, installed in such a way as to remain readily accessible and easily operable as a means to prevent danger. As such, this is a requirement for a means of switching capable of performing the functions of switching off, emergency switching and emergency stopping. It should be noted that 'readily accessible' does not necessarily require the installation of said switching device in the immediate locality of the motor being controlled.

132.15.2

4.5.4   Water heaters

Electrode heaters and boilers

These raise the temperature of the liquid being heated not through the immersion of a hot element but rather by treating the liquid as the electrolyte between electrodes. As current passes through the liquid between the electrodes its temperature is raised.

The supply to any electrode heater or boiler should be controlled by a linked circuit-breaker which, when operated, disconnects the supply to all electrodes simultaneously.

554.1.2

Water heaters having immersed, uninsulated elements

The supply to these should be via a double-pole linked switch which is either an integral part of the heating unit or placed within easy reach of it.

554.3.3

If the water heater is in a location containing a bath or shower, the relevant requirements of Section 701 regarding location and IP rating, for example, will apply.

Section 701

4.5.5   Transformers

Where a step-up transformer is installed within an installation, it is necessary to provide a linked switch on the supply side which disconnects it from all live conductors.

555.1.3

4.5.6   Outdoor (amenity) lighting installations, highway power supplies and street furniture

The requirements for these types of installation can be found in Regulation group 559.10, of which two regulations are of relevance to this publication. Regulation 559.10.6.1 allows the isolation and switching on-load of street furniture such as lighting columns in TN systems, by the withdrawal of the single-pole fuse in the supply cut-out (or multipole fuses for other equipment), providing the isolation is carried out by at least an instructed person and subject to adequate provision to prevent the equipment being unintentionally re-energized by the inadvertent replacement of the fuse. Where the electricity distributor's cut-out fuse is to be the means of isolation, the permission of the distributor must be obtained for its use as such.

559.10.6.1

559.10.6

Switching off for mechanical maintenance

5

Before considering the requirements for mechanical maintenance in BS 7671 it should be noted that where electrical equipment falls within the scope of BS EN 60204 *Safety of Machinery: Electrical equipment of machines*, the requirements for mechanical maintenance contained therein will apply.

BS EN 60204

5.1   General

BS 7671 only describes requirements for 'electrical' switching-off. It should be remembered, however, that many items of industrial equipment may involve electrically powered or actuated parts. A risk may also arise from items of equipment containing electromagnetic operation or electrical heating elements. As such it may be necessary to provide other, non-electrical, measures to prevent danger. By way of example, braking bars or similar movement-limiting devices might need to be installed to prevent the possibility of parts moving or falling. Where this is the case, any such precautions required should be clearly stated in the operational manual for that part of the installation.

Wherever mechanical maintenance work might involve a risk of physical injury, a means of switching off should be provided to facilitate such maintenance work being carried out safely.

537.3.1.1

Similar to the requirements for isolation described earlier, except where the means of switching off remains under the control of the persons dependent upon it remaining effective, it is necessary to provide a means of preventing electrically powered or operated equipment from becoming reactivated unintentionally whilst mechanical maintenance work is taking place.

537.3.1.2

5.2   Devices for switching off for mechanical maintenance

Any device provided for the purpose of switching off for mechanical maintenance should preferably be placed in the main supply circuit so as to work as directly as possible. Any switch so provided should be capable of being operated on-load at the full load current and hence be suitable for use by ordinary persons. It need not interrupt the neutral conductor.

537.3.2.1

537.3.2.5

If switching off for mechanical maintenance is achieved by interrupting the control circuit of a drive, either mechanical restraints should also be employed or the requirements of the relevant British Standard specification for the control device employed.

537.3.2.2   Whatever device is selected to provide switching off for mechanical maintenance should require manual operation. This would include a control switch acting on a contactor.

Again, as was the case for isolation, where the switching contacts cannot be seen to be in the OPEN position, a reliable means of indicating the OPEN position should be provided. This is achieved typically by 'O' and 'I' symbols indicating the OPEN (off) and CLOSED (on) positions respectively.

537.3.2.3   The means of disconnection must be so designed or installed that it cannot be switched back on inadvertently, unintentionally, or prematurely. Its purpose must be
537.3.2.4   clear either by position or by labelling and it should be placed so as to be easily used as and when required.

537.3.2.6   The use of a plug and socket-outlet arrangement is permitted for use as the means of providing switching off for mechanical maintenance where its rating does not exceed 16 A, and indeed unplugging small items is widely employed for this purpose in industrial and commercial premises. It is often reassuring for a person who is not electrically competent to see that the means of supply is so obviously disconnected.

Emergency switching and emergency stopping

6

Before considering the requirements for emergency switching in BS 7671 it should be noted that where electrical equipment falls within the scope of BS EN 60204 *Safety of Machinery: Electrical equipment of machines*, the requirements for emergency switching contained therein will apply.

BS EN 60204

6.1   General

It should be noted, as was stated in section 3.4 of this publication, that emergency switching may cause a supply to be disconnected or energized. Operation of an emergency switch might for example activate extraction plant or instigate the release of extinguishing agents. Operation of an emergency stop device might cause electromechanical braking to operate as a result of releasing a motor which was being held-off when supplied.

Emergency switching arrangements should be provided as and where required within an installation to control the electrical supply to circuits or equipment to remove a danger that has occurred unexpectedly.

537.4.1.1

In an installation supplied from a TN system, emergency switching devices need not interrupt the neutral conductor and it is preferred that they act as directly as possible on the relevant supply conductors. That is not to say, however, that emergency stopping devices cannot act upon the control circuit of, for example, a contactor.

537.4.1.2
537.4.1.3

However the means of emergency switching is arranged, its operation must not introduce other hazards or interfere with any operation initiated to remove danger. This factor needs to be borne in mind when considering the release of extinguishing agents into a switchroom for example.

537.4.1.4

6.2   Devices for emergency switching

Any device selected to act as an emergency switch should be capable of breaking the full load current likely to be present at the point of the installation where it is installed, including the current that would flow under locked rotor conditions of a motor. This requirement can lead to the need to purchase large, expensive switchgear if the emergency switching arrangement has been designed to carry the full load current, and it is for this reason that emergency stop buttons are commonly arranged, in series, to interrupt a control circuit which, when interrupted, causes a contactor to operate. Where a contactor or circuit-breaker is operated remotely, it should open when its operating coil is de-energized, although other methods providing the same degree of reliability are not excluded.

537.4.2.1

537.4.2.3

537.4.2.2 Such an arrangement is permitted by Regulation 537.4.2.2, which states that the means of emergency switching may be

▶ a single device acting directly on the supply, or
▶ a combination of equipment activated by a single action.

537.4.2.3
537.4.2.4
BS EN ISO 13850
It is preferred that emergency switching devices be of a hand-operated type, their operating button or handle being clearly identifiable by colour. If colour is used it should be red against a contrasting background. Clause 4.4.5 of BS EN ISO 13850 states that, as far as it is practicable, the background shall be coloured YELLOW. See Figure 6.1.

▼ **Figure 6.1**
Examples of emergency stop buttons [photograph courtesy of Eaton Moeller]

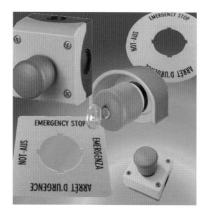

537.4.2.5 Emergency switching/stopping devices should be installed in locations where danger might occur and, where necessary, in any number of remote positions such that danger can be removed rapidly. This requirement would apply, for example, to a typical production line or an escalator.

537.4.2.6 Except for situations where the means of emergency switching and re-energizing are both under the control of the same person, the means of emergency switching should be capable of latching or otherwise being restrained in the OFF or STOP position.

537.4.2.7 As they are provided as a means of rapidly removing a source of danger, any installed emergency switching/stopping device should be clearly identifiable so that it can be easily located in an emergency and should remain readily accessible whenever the installation or part thereof controlled by the device is in use.

537.4.2.8 Regulation 537.4.2.8 states that a designer should not select a plug and socket-outlet arrangement as a means of providing emergency switching. Where there is a plug and socket-outlet in addition to the means of emergency switching, Regulation 537.4.2.8 does not preclude the use of unplugging from that socket-outlet as a means of disconnecting the supply in an emergency. If that plug and socket-outlet could be used in this way, the designer would need to consider the relative risks involved; for example, the act of unplugging could cause sparks or a dangerous emergency stopping operation.

Functional switching 7

7.1 General

Functional switching arrangements should be provided wherever it is necessary for a circuit or part thereof to be separately controlled. The type and number of switches so selected should be appropriate for the purpose. However, functional switching devices need not act upon all live conductors of a circuit. The use of the term 'act upon' here is deliberate, as functional (control) switching devices are not confined solely to two-state (i.e. on/off) switches. Dimmer switches and programmable speed controllers are examples of variable control devices.

537.5.1.1
537.5.1.3
537.5.1.2

It is quite acceptable for a single functional switching device to be used to control more than one item of equipment.

537.5.1.3

Where a functional switching device is provided to allow the changeover of supply between alternative sources, the switching device should interrupt all live conductors and should be arranged such that the sources of supply cannot operate in parallel unless they are designed to be capable of so doing.

537.5.1.4

The requirements of Regulation 537.1.2 and of Regulation 543.3.4, that combined protective earth and neutral (PEN) conductors and protective conductors respectively should not be switched, apply to functional switching.

537.1.2
543.3.4

7.2 Functional switching devices

Functional switching devices must be suitable for the most onerous duty that they will need to perform within the particular design constraints.

537.5.2.1

As mentioned above, functional switching devices need not necessarily break fully the continuity of live conductors. Although semiconductors are not permitted to act as means of isolation, they may be used for control purposes.

537.5.2.2

Clearly, any off-load devices provided within the installation for purposes such as isolation or switching for mechanical maintenance, or fuses or removable links, should not be used for functional (control) switching. In the case of switching devices and fuses this could result in damage, deterioration and/or failure. In all cases, there could well be safety issues.

537.5.2.3

7.3 Control (auxiliary) circuits

537.5.3   Control circuits should be designed, arranged and protected in such a way as to prevent the equipment being controlled from operating inadvertently in the event of a fault. It is good practice for control systems to be arranged as simply as possible while being fail-safe.

7.4 Motor control

BS EN 60204   Before considering the requirements for motor control in BS 7671 it should be noted that where electrical equipment falls within the scope of BS EN 60204 *Safety of Machinery: Electrical equipment of machines*, the requirements for mechanical maintenance contained therein will apply.

552.1.2   Every electric motor of rating greater than 0.37 kW (½ horsepower) should be provided with a control system which incorporates overload protection for the motor. This measure is not applicable to small (fractional horsepower) motors used in bathroom extractors, cooker hoods and small drives operating windows, sun-blinds and the like.

537.5.4.1
552.1.3   Except where it can be shown that a failure to restart would result in greater danger, control measures should be put in place to prevent a motor from automatically restarting after stopping as a result of voltage drop or temporary supply loss.

537.5.4.2

537.5.4.3   If, as is sometimes the case, a motor is subject to reverse-current braking, and where such reversal might result in danger, measures should be taken to prevent the reversal of direction of rotation after the driven parts come to a standstill at the end of the braking period. Further, where safety is dependent upon the motor operating in the correct direction, means should be provided to prevent reverse operation.

Firefighters' switches

8

8.1 General

As a result of the simplified structure of the 17th Edition, all requirements relating to firefighters' switches can now be found in Regulation group 537.6.

537.6

The Regulatory Reform (Fire Safety) Order 2005 (henceforth referred to as the Order) has a direct influence on many other pieces of primary and secondary legislation, requiring modifications and, in some cases, partial or full revocation of the requirements therein. It replaces fire certification under the Fire Precautions Act 1971 with a general duty to ensure, so far as is reasonably practicable, the safety of employees and a general duty, in relation to non-employees, to take such fire precautions as may reasonably be required to ensure that premises are safe.

The Order contains a number of statutory requirements relating to firefighters' switches for luminous tube signs, etc.

Article 37(4) of the Order states that a firefighters' switch must be so placed and coloured or marked as to satisfy such reasonable requirements that a fire and rescue authority may impose so that it is readily recognisable by and accessible to firefighters.

Although the detailed requirements of BS 7671 are of themselves non-statutory, article 37(5) of the Order makes it clear that 'if a firefighters' switch complies in position, colour and marking with the current edition of the IEE Wiring Regulations, fire and rescue authorities should not impose any further requirements' on such matters. It can be seen therefore that meeting the requirements of BS 7671 for the selection and installation of firefighters' switches will in itself meet the relevant statutory requirements of the relevant fire safety legislation.

BS 7671:2008 contains a number of requirements relating to firefighters' switches which are summarised below.

A firefighters' switch should be provided on the low voltage side of a circuit that supplies

537.6.1

▶ any exterior electrical installation, or
▶ an interior discharge lighting installation (not including a portable luminaire of rating not exceeding 100 W)

that operates at a voltage in excess of low voltage.

It should be noted at this point that in relation to the provision of firefighters' switches, installations in covered markets, arcades or shopping malls are considered to be exterior installations, while a temporary installation in a permanent building intended for hosting exhibitions is not.

537.6.2 Wherever practically possible, every exterior installation covered by the requirements of Regulation 537.6.1 above, within a single premises, should be controlled by a single firefighters' switch (Figure 8.1) to simplify the process of making dead all circuits therein operating at a voltage exceeding low voltage. Further, any internal installation subject to the requirements of Regulation 537.6.1 in each individual premises should be controlled by a separate single firefighters' switch, this switch being independent of a firefighters' switch for any exterior installation of the same premises.

8.2   Location and identification

537.6.3 Any firefighters' switch provided should, in the case of an exterior installation, be outside the building and be placed adjacent to the equipment it controls. Alternatively, where the switch and equipment being controlled are not adjacent to each other, notices should be placed adjacent to both the switch, indicating what it controls, and the equipment controlled, indicating the location of the appropriate switch.

For an interior installation, the switch should be placed in the main entrance of the building or other location agreed to by the local fire authority.

▼ **Figure 8.1**
Surface mounted
firefighters' switch

In all cases, the switch should be in a conspicuous position that remains accessible to firefighters. The switch should be mounted at a height of not more than 2.75 m from the ground or the standing beneath the switch.

Where more than one switch is provided for any one building, notices are required clearly describing the installation or part thereof which each switch controls.

537.6.4 Any firefighters' switch provided should be so placed to facilitate operation by a firefighter and:

▶ be coloured red
▶ be accompanied by a durable notice with the wording as shown in Figure 8.2
▶ have its ON and OFF positions clearly marked such as to be legible to a person standing in a position to operate the switch
▶ have its OFF position uppermost
▶ be so constructed such that it cannot be inadvertently switched to the ON position.

▼ **Figure 8.2**   Example of firefighter's switch label called for by Regulation 537.6.4

Note: The lettering should be at least 36 point.

Special installations or locations 9

9.1 General

The scope of BS 7671 has been extended and now contains requirements for fourteen types of special installation or location. These requirements add to, replace or modify those contained within Parts 1 to 6 of the Regulations. Of these fourteen sections, seven contain specific requirements falling within the remit of this publication. These are summarised below.

110.1

Part 7

9.2 Construction sites

Section 704 gives requirements for construction and demolition site installations. Regulation 704.537.2.2 requires that each Assembly for Construction Sites (ACS) incorporates suitable means of isolating the incoming supply to the temporary installation.

Section 704

704.537.2.2

Other standards pertinent to construction sites are:

▶ BS 4363:1998 (2007) *Distribution assemblies for reduced low voltage electricity supplies for construction and building sites*
▶ BS 7375:1996 *Code of practice for distribution of electricity on construction and building sites*
▶ BS EN 60439-4:2004 *Low voltage switchgear and controlgear assemblies. Particular requirements for assemblies for construction sites.*

BS 4363

BS 7375

BS EN 60439-4

Clause 4.1.1 of BS 7375 requires that any devices intended to be used as a means of isolation should be of the multipole type, arranged to break all circuit conductors (including the neutral conductor) simultaneously. This means of isolation should be capable of being secured in the OFF position by means of a key or special tool.

Clause 101.2 of BS EN 60439-4 requires any incoming unit to include a means of isolation that can be secured in the OPEN position.

9.3 Agricultural and horticultural premises

It is a requirement for any electrical heating appliance selected for use within an agricultural or horticultural installation to incorporate a visual indication of whether it is in the ON or OFF position.

Section 705

705.53

It remains a requirement for the electrical installation of each building or part of a building to be provided with a separate means of isolation.

705.537.2

Some items of electrical equipment in agricultural and horticultural premises remain subject to seasonal patterns of usage connected for example to harvesting times. Where this is the case, a means of isolation acting on all live conductors, including the neutral conductor, should be provided.

Isolation and switching devices installed within agricultural or horticultural premises should be so placed as to be out of the reach of livestock to prevent damage or unintended operation and such that they will not be made inaccessible when required. Furthermore, a clear indication should be given of the purpose of these isolating devices where this is not immediately apparent.

9.4 Marinas

Section 709

The 17th Edition contains requirements applicable to electrical installations in marinas.

709.537.2.1.1
709.411.4 Note

709.537.2.1.1

It is commonplace in marina installations for distribution cabinets to be provided containing the equipment necessary for the connection of a number of vessels. Regulation 709.537.2.1.1 requires that each distribution cabinet should be provided with a means of isolation. The Electricity Safety, Quality and Continuity Regulations 2002 effectively prohibit the use of a TN-C-S supply to be taken to boats. Isolators installed in marina installations must disconnect all live conductors including the neutral conductor. It is permitted for one isolator to serve up to four socket-outlets (see Figures 9.1 and 9.2).

9.5 Exhibitions, shows and stands

Section 711

Another type of installation newly included in the Wiring Regulations relates to exhibitions, shows and stands.

711.537.2.3

Regulation 711.537.2.3 requires that each separate temporary structure, vehicle, stand or unit which is intended to be occupied by persons and each distribution circuit being used to provide a supply to outdoor installations of this type should be provided with its own means of isolation. The means of isolation should be readily accessible and clearly identifiable.

711.55.4.1

Where electric motors are provided and a hazardous situation might develop, a means of isolation acting on all poles of its supply should be provided adjacent to the motor.

711.559.4.7

Where the installation of an exhibition, show or stand contains discharge lighting operating at a voltage in excess of 230/400 V a.c. these should be supplied from a separate circuit which is controlled by at least one emergency switch. The emergency switches should be easily visible, accessible and clearly identified.

9.6 Solar photovoltaic (PV) power supply systems

Section 712

Section 712 contains requirements for this type of installation.

551.7.6

For a solar PV installation designed to operate in parallel with the public supply, reference should be made to Regulation 551.7.6 and the guidance on this regulation given in section 4.5.2 of this publication.

▼ Figure 9.1 Example of the means of connection for use with vessels in marinas (single-phase)

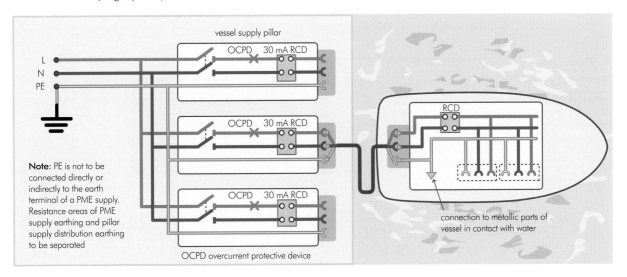

▼ Figure 9.2 Example of the means of connection for use with vessels in marinas (three-phase)

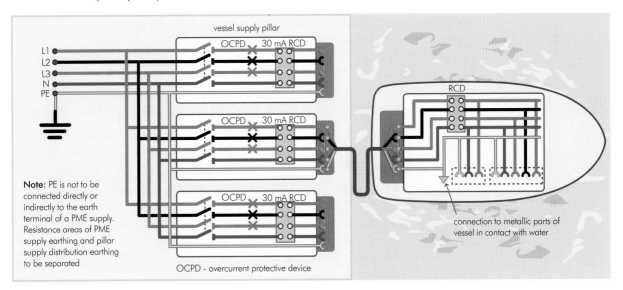

There are particular problems associated with solar PV power supply systems in that it is difficult to remove all sources of light acting upon the solar PV arrays (solar panels). This is recognised in Regulation 712.410.3, which states that PV equipment on the d.c. side shall be considered to be energized, even when the system is disconnected from the a.c. side.

712.410.3

As a means of facilitating maintenance of the PV convertor (invertor), isolation devices must be provided to allow the PV convertor to be disconnected from both the d.c. side and the a.c. side. This arrangement is illustrated in Figure 712.1 of BS 7671. It should be noted at this point that Regulation 712.537.2.2.5 also requires the installation of a switch-disconnector to be installed on the supply side of the d.c. convertor.

712.537.2.1.1

Figure 712.1
712.537.2.2.5

When considering the selection and erection of devices for isolation and switching for use between the solar PV supply system and the public supply network, the public supply should be considered to be the source (as defined) and the PV installation should be considered to be the load.

712.537.2.2.1

712.537.2.2.5.1 As a result of the difficulties involved in confirming that solar PV systems are fully isolated as mentioned above, it is a requirement to post warning notices on all junction boxes employed within the solar PV system, pointing out that parts within the boxes might still be live after isolation from the PV convertor (Figure 9.3).

▼ **Figure 9.3**
Example of warning label required for junction boxes in solar PV systems

WARNING
PV SYSTEM
Parts inside this box or enclosure may still be live after isolation from the supply.

9.7   Caravans and motor caravans

Section 721   Section 721 contains particular requirements to be applied to electrical installations in caravans and motor caravans.

721.537.2.1.1   Each individual caravan or motor caravan should be provided with a main isolator arranged to disconnect all live conductors. It should be placed to allow for ready operation. If the caravan consists of only one final circuit, the overcurrent protective device for the circuit may be used as the isolator.

721.537.2.1.1.1
Figure 721   Regulation 721.537.2.1.1.1 requires the installation of an advisory notice close to the main isolating switch. The wording required for this notice is given in Figure 721. This includes an explanation of how to disconnect the caravan making use of the main isolating switch.

9.8   Structures, amusement devices and booths at fairgrounds, amusement parks and circuses

Section 740   Section 740 is a new section applicable to structures, amusement devices and booths at fairgrounds and the like.

740.537.1   Such installations are likely to comprise a number of separate structures, booths, rides or similar. Regulation 740.537.1 requires each booth, structure, ride or similar to be provided with its own means of isolation which should remain readily accessible at all times. A separate means of isolation is also required for each distribution circuit intended to supply outdoor installations.

740.537.2.1.1

740.537.2.2   All switching devices selected to act as the means of isolation in such locations should disconnect all live conductors including the neutral conductor.

740.55.3.2   Any luminous tubes or signs which operate at a voltage exceeding 230/400 V a.c. should be connected to a circuit which is controlled by an emergency switching arrangement (see Chapter 8).

Safety service and product standards of relevance to this publication

A

BS or EN number	Title	References in this Guidance Note
BS 88 series	Low-voltage fuses	Table 3.1
BS 1362:1973	Specification for general purpose fuse links for domestic and similar purposes (primarily for use in plugs)	Table 3.1
BS 1363 series	13 A plugs, socket-outlets, adaptors and connection units	3.1
BS 1363-1:1995	Specification for rewirable and non-rewirable 13 A fused plugs	Table 3.1
BS 1363-2:1995	Specification for 13 A switched and unswitched socket-outlets	Table 3.1
BS 1363-4:1995	Specification for 13 A fused connection units switched and unswitched	Table 3.1
BS 3676-1:2000	Switches for household and similar fixed electrical installations. General requirements	3.1; Table 3.1
BS 4177:1992	Specification for cooker control units	Table 3.1
BS 4363:1998 (2007)	Specification for distribution assemblies for reduced low voltage electricity supplies for construction and building sites	9.2
BS 4884 series	Technical manuals	Introduction
BS 4940 series	Technical information on constructional products and services	Introduction
BS 5266-1:2005	Emergency lighting. Code of practice for the emergency lighting of premises	4.3
BS 5839 series	Fire detection and fire alarm systems for buildings	
BS 5839-1:2002 (2008)	Code of practice for system design, installation, commissioning and maintenance	4.3
BS 5839-6:2004	Code of practice for the design, installation and maintenance of fire detection and fire alarm systems in dwellings	4.3
BS 5518:1997	Specification for electronic variable control switches (dimmer switches) for tungsten filament lighting	3.1
BS 5733:1995	Specification for general requirements for electrical accessories	3.1; Table 3.1
BS 6972:1988	Specification for general requirements for luminaire supporting couplers for domestic, light industrial and commercial use	Table 3.1
BS 7375:1996	Code of practice for distribution of electricity on construction and building sites	9.2
BS 7671:2008	Requirements for electrical installations. IEE Wiring Regulations. Seventeenth Edition	Numerous

BS or EN number	Title	References in this Guidance Note
BS EN 60073:2002	Basic and safety principles for man-machine interface, marking and identification. Coding principles for indicators and actuators	3.1
BS EN 60204 series	Safety of machinery. Electrical equipment of machines	1.2.4; 5; 6; 7.4
BS EN 60204-1:2006 (2009)	Electrical equipment of machines. General requirements	3.5
BS EN 60309 series	Plugs, socket-outlets and couplers for industrial purposes	Table 3.1
BS EN 60309-2:1999	Dimensional interchangeability requirements for pin and contact-tube accessories	3.1
BS EN 60439-4:2004	Low-voltage switchgear and controlgear assemblies. Particular requirements for assemblies for construction sites (ACS)	9.2
BS EN 60669 series	Switches for household and similar fixed electrical installations	3.1
BS EN 60669-1:2000 (2008)	General requirements (also known as BS 3676-1:2000)	Table 3.1
BS EN 60669-2-1:2004	Particular requirements. Electronic switches	Table 3.1
BS EN 60669-2-2:2006	Particular requirements. Electromagnetic remote-control switches (RCS)	Table 3.1
BS EN 60669-2-3:1999	Particular requirements. Time-delay switches (TDS)	Table 3.1
BS EN 60669-2-4	Particular requirements. Isolating switches	Table 3.1
BS EN 60898 series	Circuit breakers for overcurrent protection for household and similar installations	Table 3.1
BS EN 60898-1:2003	Circuit breakers for a.c. operation	Table 3.1
BS EN 60898-2:2006	Circuit breakers for a.c. and d.c. operation	Table 3.1
BS EN 60947 series	Low-voltage switchgear and controlgear	3.1
BS EN 60947-1:2007	General rules	3.8
BS EN 60947-2:2006	Circuit-breakers	Table 3.1
BS EN 60947-3:1999	Switches, disconnectors, switch-disconnectors and fuse-combination units	Table 3.1
BS EN 60947-4-1:2001	Contactors and motor-starters. Electromechanical contactors and motor-starters	Table 3.1
BS EN 60947-5-1:2004	Control circuit devices and switching elements. Electromechanical control circuit devices	Table 3.1
BS EN 60947-5-5:1998	Control circuit devices and switching elements. Electrical emergency stop devices with mechanical latching function	3.5
BS EN 60947-6-1:2005	Multiple function equipment. Transfer switching equipment	Table 3.1
BS EN 60947-6-2:2003	Multiple function equipment. Control and protective switching devices (or equipment) (CPS)	Table 3.1
BS EN 61008-1:2004	Residual current operated circuit-breakers without integral overcurrent protection for household and similar uses (RCCBs). General rules	Table 3.1
BS EN 61009-1:2004	Residual current operated circuit-breakers with integral overcurrent protection for household and similar uses (RCBOs). General rules	Table 3.1
BS EN 61095:2009	Electromechanical contactors for household and similar purposes	Table 3.1

BS or EN number	Title	References in this Guidance Note
BS EN ISO 12100 series	Safety of machinery. Basic concepts, general principles for design	1.2.4
BS EN ISO 13850:2008	Safety of machinery. Emergency stop equipment, functional aspects. Principles for design	1.2.4; 3.5
BS IEC 61995-1:2008	Devices for the connection of luminaires for household and similar purposes. General requirements	Table 3.1
IEC 60050	International electrotechnical vocabulary. Multiple (110) part standard	2; 3.8
IEC 60884 series	Plugs and socket-outlets for household and similar purposes	Table 3.1
IEC 60906 series	IEC system of plugs and socket-outlets for household and similar purposes	Table 3.1

Safe isolation procedures

This appendix is not intended to be an exhaustive treatment of the subject of safe isolation, but rather a reminder of the minimum steps which should be taken to confirm that an installation, circuit or item of equipment has been made dead. References are given at the end of this appendix for further sources of information regarding safe isolation.

Minimum stages to confirm safe isolation:

- Locate/positively identify correct isolation point or device
- Check condition of voltage indicating device
- Confirm that voltage indicating device is functioning correctly
- Switch off installation/circuit to be isolated
- Verify with voltage indicating device that no voltage is present
- Re-confirm that voltage indicating device functions correctly on known supply or proving unit
- Lock off or otherwise secure device used to isolate installation/circuit
- Post warning notice(s).

In order to identify appropriate points or devices to carry out the isolation, it is important to employ all relevant sources of information. These would include discussions with the persons responsible for the electrical installation and perusal of building operation manuals, installation schematic drawings and circuit schedules.

It is highly recommended that a voltage indicator or test lamp which has been expressly designed for proving the absence of supply is employed when carrying out safe isolation. These have a number of design features to protect the user including integral current-limiting measures; finger guards; appropriately insulated leads; and probes arranged such that a minimum (2 to 4 mm) of uninsulated tip is exposed. The use of multi-range instruments is not recommended. They may lack a number of the safety features of a voltage indicator mentioned above, and incorrect ranges may be selected, both of which may place a user in danger.

The voltage indicator/test lamp should be checked for signs of damage or deterioration prior to each use. Any voltage indicator/test lamp with significant damage resulting for example in exposed conductive parts which will be live during use should be removed from service immediately and if not repairable be destroyed to prevent its unintended or inadvertent use.

It is essential that the user confirms the functionality of the voltage indicator/test lamp before EACH use. This may be achieved by proving on a known supply or proprietary proving unit, even where the voltage indicator has an integral 'self-test' facility.

After switching off the installation, circuit or item of equipment to be isolated, the 'proved' voltage indicator/test lamp should be used to confirm the absence of any voltage. The correct operation of the voltage indicator/test lamp should then be reconfirmed using a known supply or proving unit.

Having confirmed that the correct installation, circuit or item of equipment has been made dead, the means of isolation should be made secure. Unless the means of isolation remains under the control of the person(s) dependent upon it, this will involve the use of some means of locking off. A wide range of lock-out devices are available allowing virtually any type of device (circuit-breaker, switch, plug, etc.) to be secured against inadvertent or unauthorised use. It is never acceptable to merely place a strip of tape over a device which has been switched off.

Any warning notice posted should convey in simple terms that the installation, circuit or item of equipment has been deliberately disconnected from the supply (see Figure B.1). It should inform the reader that care must be taken to ensure the safety of persons who are reliant upon the continued effectiveness of the isolation and should state that the installation, circuit or equipment should not be re-energized until it has been confirmed that it is safe to do so.

▼ **Figure B.1**
Examples of safe
isolation warning notice

DO NOT ATTEMPT TO REMOVE

THIS SOURCE OF ELECTRICAL SUPPLY HAS BEEN DELIBERATELY ISOLATED AND SECURED.

Electrical **excellence**   www.**theiet**.org/wiringregs

MINIMUM STAGES FOR SAFE ISOLATION

- Locate/positively identify correct isolation point or device
- Check condition of voltage indicating device
- Confirm that voltage indicating device is functioning correctly
- Switch off installation/circuit to be isolated
- Verify with voltage indicating device that no voltage is present
- Re-confirm that voltage indicating device functions correctly on known supply/proving unit
- Lock-off or otherwise secure device used to isolate installation /circuit
- Post warning notice(s)

MINIMUM STAGES FOR SAFE ISOLATION
Electrical **excellence**

- Locate/positively identify correct isolation point or device
- Check condition of voltage indicating device
- Confirm that voltage indicating device is functioning correctly
- Switch off installation/circuit to be isolated
- Verify with voltage indicating device that no voltage is present
- Re-confirm that voltage indicating device functions correctly on known supply/proving unit
- Lock-off or otherwise secure device used to isolate installation/circuit
- Post warning notice(s)   www.**theiet**.org/wiringregs

The legal requirements relating to safe isolation procedures can be found in Regulation 12 'Means for cutting off the supply and for isolation' and Regulation 13 'Precautions for work on equipment made dead' of the Electricity at Work Regulations (EWR) 1989.

Employers, employees and the self-employed working on or near electrical installations, or responsible for such work, should be familiar with the statutory requirements imposed on them by the EWR 1989.

The Health and Safety Executive (HSE) has produced the *Memorandum of guidance on the Electricity at Work Regulations 1989* (HSR25) to clarify the requirements of the EWR 1989. It is essential reading for all persons upon whom the EWR 1989 impose duties.

Another HSE publication *Electrical test equipment for use by electricians* (GS 38) gives guidance on the selection and use of voltage indicating devices to be used when carrying out safe isolation procedures. This too is essential reading.

The HSE publication *Electricity at work. Safe working practices* (HSG85) gives a broader overview of all aspects of working on or near electrical installations.

Other valuable sources of information on this matter can be found online on the following links:

▶ SELECT/HSE guide on safe isolation procedures: www.select.org.uk/downloads/publications/Select%20-%20Safe%20Isolation%20Procedures.pdf
▶ ESC Best Practice Guide *Guidance on safe isolation procedures for low voltage installations*: http://niceic.org.uk/inc/file-get.asp?FILE=BPG2_March07(sec).pdf
▶ Elliott, J. 'Safe isolation of low voltage installations'. *IET Wiring Matters*. Issue 30, Spring 2009, pp. 12–17: www.theiet.org/publishing/wiring-regulations/mag/index.cfm

Permit-to-work procedures C

A permit-to-work is a formal procedure designed to make employees aware that essential precautions and, where necessary, physical safeguards such as carrying out safe isolation, locking off and installation of mechanical constraints, have been put in place.

Permits-to-work are issued as part of a *safe system of work*. Generally, a permit-to-work is a management procedure in which only persons having specific management authority can sign a permit allowing activities, such as in the case of electrical installations, safe isolation or switching off for mechanical maintenance, upon which a person's life might depend, to be carried out. A permit-to-work is effectively a statement that measures have been taken such that it is safe to work. As such, a permit-to-work should not be issued until after an installation, circuit, or item of equipment has been safely isolated and, in the case of switching off for mechanical maintenance, all other precautions required to prevent movement of electrically actuated parts have been put in place.

An electrical permit-to-work is primarily a statement that a circuit or item of equipment is safe to work on. A permit-to-work should never be issued on equipment that is still live.

A permit-to-work should be kept as simple as possible, contain clear, concise instructions including simple circuit diagrams or schematic drawings to be followed and should, in the case of a failure to properly implement, be supported by the implementation of appropriate disciplinary measures.

A permit-to-work should include the following:

▶ The person(s) to whom the permit applies (the person(s) who will work on the equipment). In practice one individual may be designated as the responsible person
▶ Identification of the circuit(s) or equipment which has been made dead and its precise location
▶ The points used for isolation (switchgear location and/or reference number)
▶ Where conductors are earthed, where applicable
▶ Location where safety locks have been applied and warning notices have been posted
▶ The nature of the work activity to be carried out
▶ The presence of any other sources of hazard, with a cross-reference to other relevant permits which may have been issued
▶ Details of any other precautions which have been taken during the course of the work.

The person issuing the permit should explain the extent of the work covered to the person receiving it. Both persons should agree the extent of the work covered by the permit before they sign it.

The subject of permits-to-work is covered in greater detail in the HSE publication *Electricity at work. Safe working practices* (HSG85) and is also discussed in the IET publication *Electrical Maintenance*.

A model permit-to-work form and associated guidance is included on the following pages.

Notes on Model Form of Permit-to-Work

1   Access to and work in fire protected areas

Automatic control

Unless alternative approved procedures apply because of special circumstances then before access to, or work or other activities are carried out in, any enclosure protected by automatic fire extinguishing equipment:

a   the automatic control shall be rendered inoperative and the equipment left on hand control. A caution notice shall be attached.

b   precautions taken to render the automatic control inoperative and the conditions under which it may be restored shall be noted on any safety document or written instruction issued for access, work or other activity in the protected enclosure.

c   the automatic control shall be restored immediately after the persons engaged on the work or other activity have withdrawn from the protected enclosure.

2   Procedure for issue and receipt

a   A permit-to-work shall be explained and issued to the person in direct charge of the work, who after reading its contents to the person issuing it, and confirming that he/she understands it and is conversant with the nature and extent of the work to be done, shall sign its receipt and its duplicate.

b   The recipient of a permit-to-work shall be a competent person who shall retain the permit-to-work in his/her possession at all times whilst work is being carried out.

c   Where more than one working party is involved, a permit-to-work shall be issued to the competent person in direct charge of each working party and these shall, where necessary, be cross-referenced one with another.

3   Procedure for clearance and cancellation

a   A permit-to-work shall be cleared and cancelled:
  i   when work on the apparatus or conductor for which it was issued has been completed;
  ii   when it is necessary to change the person in charge of the work detailed on the permit-to-work;
  iii   at the discretion of the responsible person when it is necessary to interrupt or suspend the work detailed on the permit-to-work.

b   The recipient shall sign the clearance and return to the responsible person who shall cancel it. In all cases the recipient shall indicate in the clearance section whether the work is 'complete' or 'incomplete' and that all gear and tools 'have' or 'have not' been removed.

c   Where more than one permit-to-work has been issued for work on apparatus or conductors associated with the same circuit main earths, the controlling engineer shall ensure that all such permits-to-work have been cancelled before the circuit main earths are removed.

4   Procedure for temporary withdrawal or suspension

Where there is a requirement for a permit-to-work to be temporarily withdrawn or suspended this shall be in accordance with an approved procedure.

PERMIT-TO-WORK (front)

1. ISSUE

No. ……………..

To ……………………………………………………………………………………………………..

The following apparatus has been made safe in accordance with the safety rules
for the work detailed on this permit-to-work to proceed:

…………………………………………………………………………...……………………..

…………………………………………………………………………...……………………..

Treat all other apparatus as live.

Circuit main earths are applied at:

…………………………………………………………………………...……………………..

…………………………………………………………………………...……………………..

Other precautions and information and any local instructions applicable to
the work (Notes 1 and 2):

…………………………………………………………………………...……………………..

…………………………………………………………………………...……………………..

The following work is to be carried out:

…………………………………………………………………………...……………………..

…………………………………………………………………………...……………………..

…………………………………………………………………………...……………………..

Name (block capitals) ………………………………………………………………………

Signature ………………………………….. Time …………………….. Date ……………………..

5. DIAGRAM (see over for sections 2, 3 and 4 of this permit)

The diagram should show:
  a the safe zone where work is to be carried out
  b the points of isolation
  c the places where earths have been applied, and
  d the locations where 'danger' notices have been posted.

PERMIT-TO-WORK (back)

2. RECEIPT
(Note 2)

I accept responsibility for carrying out the work on the apparatus detailed on this permit-to-work and no attempt will be made by me, or by the persons under my charge, to work on any other apparatus.

Name (block capitals) ....

Signature .... Time .... Date ....

3. CLEARANCE
(Note 3)

All persons under my charge have been withdrawn and warned that it is no longer safe to work on the apparatus detailed on this permit-to-work, and all additional earths have been removed.

The work is complete* / incomplete*

All gear and tools have* / have not* been removed

Name (block capitals) ....

Signature .... Time .... Date ....

*Delete words not applicable.

4. CANCELLATION
(Note 3)

This permit-to-work is cancelled.

Name (block capitals) ....

Signature .... Time .... Date ....

5. DIAGRAM (continue below if needed)

Index

The

IEE Wiring Regulations and associated publications

The IEE prepares regulations for the safety of electrical installations for buildings, the *IEE Wiring Regulations* (BS 7671 *Requirements for Electrical Installations*), which have now become the standard for the UK and many other countries. It also recommends, internationally, the requirements for ships and offshore installations. The IEE provides guidance on the application of the installation regulations through publications focused on the various activities from design of the installation through to final test and then maintenance. This includes a series of eight Guidance Notes, two Codes of Practice and Model Forms for use in Wiring Installations.

Requirements for Electrical Installations BS 7671:2008 (IEE Wiring Regulations, 17th Edition)
Order book PWR1700B Paperback 2008
ISBN: 978-0-86341-844-0   **£75**

On-Site Guide (BS 7671:2008 17th Edition)
Order book PWGO170B 188pp Paperback 2008
ISBN: 978-0-86341-854-9   **£22**

Wiring Matters Magazine   FREE
If you wish to receive a FREE copy or advertise in Wiring Matters please visit
www.theiet.org/wm

IEE Guidance Notes

A series of Guidance Notes has been issued, each of which enlarges upon and amplifies the particular requirements of a part of the IEE Wiring Regulations.

Guidance Note 1: Selection & Erection of Equipment, 5th Edition
Order book PWG1170B 216pp Paperback 2009
ISBN: 978-0-86341-855-6   **£30**

Guidance Note 2: Isolation & Switching, 5th Edition
Order book PWG2170B 88pp Paperback 2009
ISBN: 978-0-86341-856-3   **£25**

Guidance Note 3: Inspection & Testing, 5th Edition
Order book PWG3170B 128pp Paperback 2008
ISBN: 978-0-86341-857-0   **£25**

Guidance Note 4: Protection Against Fire, 5th Edition
Order book PWG4170B 104pp Paperback 2009
ISBN: 978-0-86341-858-7   **£25**

Guidance Note 5: Protection Against Electric Shock, 5th Edition
Order book PWG5170B 144pp Paperback 2009
ISBN: 978-0-86341-859-4   **£25**

Guidance Note 6: Protection Against Overcurrent, 5th Edition
Order book PWG6170B 104pp Paperback 2009
ISBN: 978-0-86341-860-0   **£25**

Guidance Note 7: Special Locations, 3rd Edition
Order book PWG7170B 144pp Paperback 2009
ISBN: 978-0-86341-861-7   **£25**

Guidance Note 8: Earthing & Bonding, 1st Edition
Order book PWRG0241 168pp Paperback 2007
ISBN: 978-0-86341-616-3   **£25**

continues overleaf ▶

Other guidance publications

**Commentary on IEE Wiring Regulations
(17th Edition, BS 7671:2008)**
Order book PWR08640
c.432pp Hardback 2009
ISBN: 978-0-86341-966-9   **£65**

Electrical Maintenance, 2nd Edition
Order book PWR05100
228pp Paperback 2006
ISBN: 978-0-86341-563-0   **£40**

**Code of Practice for In-service Inspection and
Testing of Electrical Equipment, 3rd Edition**
Order book PWR08630
152pp Paperback 2007
ISBN: 978-0-86341-833-4   **£40**

**Electrical Craft Principles, Volume 1,
5th Edition**
Order book PBNS0330
344pp Paperback 2009
ISBN: 978-0-86341-932-4   **£25**

**Electrical Craft Principles, Volume 2,
5th Edition**
Order book PBNS0340
432pp Paperback 2009
ISBN: 978-0-86341-933-1   **£25**

**Electrician's Guide to the Building
Regulations, 2nd Edition**
Order book PWGP170B
234pp Paperback 2008
ISBN: 978-0-86341-862-4   **£22**

**Electrical Installation Design Guide:
Calculations for Electricians and Designers**
Order book PWR05030
186pp Paperback 2008
ISBN: 978-0-86341-550-0   **£22**

Electrician's Guide to Emergency Lighting
Order book PWR05020
88pp Paperback 2009
ISBN: 978-0-86341-551-7   **£22**

Electrical training courses

We offer a comprehensive range of technical training at
many levels, serving your training and career development
requirements as and when they arise.

Courses range from Electrical Basics to Qualifying City &
Guilds or EAL awards.

Train to the 17th Edition BS 7671:2008
▶ Update from 16th to 17th Edition
▶ Understand the changes
▶ New qualifying awards C&G/EAL
▶ Meet industry standards

Qualifying Courses
▶ Certificate of Competence Management of Electrical
  Equipment Maintenance (PAT) – 1 day
▶ Certificate of Competence for the Inspection and
  Testing of Electrical Equipment (PAT) – 1 day
▶ Certificate in the Requirements for Electrical
  Installations – 3 days
▶ Upgrade from 16th Edition achieved since 2001 –
  1 day
▶ Certificate in Fundamental Inspection, Testing and
  Internal Verification – 3 days
▶ Certificate in Inspection, Testing and Certification of
  Electrical Installations – 3 days

Other 17th Edition Courses
▶ Earthing & Bonding – For designers and electrical
  contractors who require a good working knowledge
  of the E & B arrangements as required by
  BS 7671:2008
▶ 17th Edition Design – BS 7671 and the principles
  associated with the design of electrical installations

To view all our current courses and book online, visit
www.theiet.org/coursesbr

**To discuss your training requirements and for on-
site group training, please speak to one of our
advisors on +44 (0)1438 767289**

Collective **inspiration**

Order Form

How to order

BY PHONE:
+44 (0)1438 767328
BY FAX:
+44 (0)1438 767375
BY EMAIL:
sales@theiet.org
BY POST:
The Institution of
Engineering
and Technology,
PO Box 96,
Stevenage
SG1 2SD, UK
OVER THE WEB:
www.theiet.org/books

*Postage/Handling: Postage within the UK is £3.50 for any number of titles. Outside UK (Europe) add £5.00 for first title and £2.00 for each additional book. Rest of World add £7.50 for the first book and £2.00 for each additional book. Books will be sent via airmail. Courier rates are available on request, please call +44 (0) 1438 767328 or email sales@theiet.org for rates.

** To qualify for discounts, member orders must be placed directly with the IET.

GUARANTEED RIGHT OF RETURN:
If at all unsatisfied, you may return book(s) in new condition within 30 days for a full refund. Please include a copy of the invoice.

DATA PROTECTION:
The information that you provide to the IET will be used to ensure we provide you with products and services that best meet your needs. This may include the promotion of specific IET products and services by post and/or electronic means. By providing us with your email address and/or mobile telephone number you agree that we may contact you by electronic means. You can change this preference at any time by visiting www.theiet.org/my.

Details

Name:

Job Title:

Company/Institution:

Address:

Postcode: Country:

Tel: Fax:

Email:

Membership No (if Institution member):

Payment methods

☐ By **cheque** made payable to The Institution of Engineering and Technology

☐ By **credit/debit card:**

☐ Visa ☐ Mastercard ☐ American Express ☐ Maestro Issue No:_____

Valid from: ☐☐ ☐☐   Expiry Date: ☐☐ ☐☐   Card Security Code: ☐☐☐☐
(3 or 4 digits on reverse of card)

Card No: ☐☐☐☐ ☐☐☐☐ ☐☐☐☐ ☐☐☐☐

Signature_____ Date _____
(Orders not valid unless signed)

Cardholder Name:

Cardholder Address:

Town: Postcode:

Country:

☐ By official **company purchase order** (please attach copy)
EU VAT number:_____

Ordering information

Quantity	Book No.	Title/Author	Price (£)
		Subtotal	
		- Member discount**	
		+ Postage /Handling*	
		+ VAT (if applicable)	
		Total	

Membership

Passionate about engineering? Committed to your career?

Do you want to join an organisation that is inspiring, insightful and innovative?

One of the most highly recognised knowledge sharing networks in the world, membership to the Institution of Engineering and Technology (IET) is for engineers and technologists working or studying in an increasingly multidisciplined, digital and global environment.

Joining the IET and having access to tailored products and services will become invaluable for your career and can be your first step towards professional qualifications.

You could take advantage of ...

▶ 18 issues per year of the industry's leading publication, *E&T* magazine.

▶ Professional development and career support services to help gain professional registration.

▶ Discounted rates on dedicated training courses, seminars and events covering a wide range of subjects and skills.

▶ Watch live IET.tv event footage at your desktop via the internet, ask the speaker questions during live streaming and feel part of the audience without physically being there.

▶ Access to over 100 local networks around the world.

▶ Meet like-minded professionals through our array of specialist online communities.

▶ Instant online access to over 70,000 books, 3,000 periodicals and full-text collections of electronic articles – wherever you are in the world.

▶ Discounted rates on IET books and technical proceedings.

Join online today at www.theiet.org/join or contact our membership and customer service centre on +44 (0)1438 765678

Professional Registration

What type of registration is for you?

Engineering Technicians (EngTech) are involved in applying proven techniques and procedures to the solution of practical engineering problems. You will carry supervisory or technical responsibility, and are competent to exercise creative aptitudes and skills within defined fields of technology. Engineering Technicians also contribute to the design, development, manufacture, commissioning, operation or maintenance of products, equipment, processes or services.

Incorporated Engineers (IEng) maintain and manage applications of current and developing technology, and may undertake engineering design, development, manufacture, construction and operation. Incorporated Engineers are engaged in technical and commercial management and possess effective interpersonal skills.

Chartered Engineers (CEng) develop appropriate solutions to engineering problems, using new or existing technologies, through innovation, creativity and change. They might develop and apply new technologies, promote advanced designs and design methods, introduce new and more efficient production techniques, marketing and construction concepts, pioneer new engineering services and management methods. Chartered Engineers are engaged in technical and commercial leadership and possess interpersonal skills.

For further information on Professional Registration (EngTech/IEng/CEng), tel: +44 (0)1438 765673 or email: membership@theiet.org

Notes

Notes

Notes

Notes